**Wide Bandgap Light
Emitting Materials
and Devices**

*Edited by
Gertrude F. Neumark,
Igor L. Kuskovsky, and
Hongxing Jiang*

1807–2007 Knowledge for Generations

Each generation has its unique needs and aspirations. When Charles Wiley first opened his small printing shop in lower Manhattan in 1807, it was a generation of boundless potential searching for an identity. And we were there, helping to define a new American literary tradition. Over half a century later, in the midst of the Second Industrial Revolution, it was a generation focused on building the future. Once again, we were there, supplying the critical scientific, technical, and engineering knowledge that helped frame the world. Throughout the 20th Century, and into the new millennium, nations began to reach out beyond their own borders and a new international community was born. Wiley was there, expanding its operations around the world to enable a global exchange of ideas, opinions, and know-how.

For 200 years, Wiley has been an integral part of each generation's journey, enabling the flow of information and understanding necessary to meet their needs and fulfill their aspirations. Today, bold new technologies are changing the way we live and learn. Wiley will be there, providing you the must-have knowledge you need to imagine new worlds, new possibilities, and new opportunities.

Generations come and go, but you can always count to Wiley to provide you the knowledge you need, when and where you need it!

William J. Pesce
President and Chief Executive Officer

Peter Booth Wiley
Chairman of the Board

Wide Bandgap Light Emitting Materials and Devices

Edited by
Gertrude F. Neumark, Igor L. Kuskovsky,
and Hongxing Jiang

WILEY-VCH Verlag GmbH & Co. KGaA

The Editors

Prof. Dr. Gertrude F. Neumark
Dept. of Applied Physics
Columbia University
1137, 500 W. 120th Street
New York, NY 10027
USA

Dr. Igor Kuskovsky
Department of Physics
Queens College of CUNY
65-30 Kissena Blvd.
Flushing, NY 11367
USA

Prof. Dr. Hongxing Jiang
Kansas State University
Department of Physics
Cardwell Hall
Manhattan, KS 66506-2601
USA

Cover
Courtesy of Andrei Osinsky, see also article by
John Muth and Andrei Osinsky in this book.

Wiley Bicentennial Logo
Richard J. Pacifico

All books published by Wiley-VCH are carefully
produced. Nevertheless, authors, editors, and
publisher do not warrant the information contained
in these books, including this book, to be free of
errors. Readers are advised to keep in mind that
statements, data, illustrations, procedural details or
other items may inadvertently be inaccurate.

Library of Congress Card No.:
applied for

British Library Cataloguing-in-Publication Data
A catalogue record for this book is available from
the British Library.

**Bibliographic information published by the Deutsche
Nationalbibliothek**
The Deutsche Nationalbibliothek lists this
publication in the Deutsche Nationalbibliografie;
detailed bibliographic data are available in the
Internet at <http://dnb.d-nb.de>.

© 2007 WILEY-VCH Verlag GmbH & Co. KGaA,
Weinheim

All rights reserved (including those of translation
into other languages). No part of this book may be
reproduced in any form – by photoprinting,
microfilm, or any other means – nor transmitted or
translated into a machine language without written
permission from the publishers. Registered names,
trademarks, etc. used in this book, even when not
specifically marked as such, are not to be
considered unprotected by law.

Composition SNP Best-set Typesetter Ltd.,
Hong Kong

Printing betz-druck GmbH, Darmstadt

Bookbinding Litges & Dorf GmbH, Heppenheim

Printed in the Federal Republic of Germany
Printed on acid-free paper

ISBN 978-3-527-40331-8

QC611
.8
W53 W525
2007
PHYS

Contents

Wide Bandgap Light Emitting Materials and Devices. Edited by G. F. Neumark, I. L. Kuskovsky, and H. Jiang
Copyright © 2007 WILEY-VCH Verlag GmbH & Co. KGaA, Weinheim
ISBN: 978-3-527-40331-8

Preface

The most recent era of progress in semiconductor light emitting devices and materials started around 1990, with two independent developments. The first, in 1991 was a report by Haase et al. (Appl. Phys. Lett. **59** (1991) 1272) of the first blue-green laser diode, made from ZnSe and related alloys. The second, in 1994, was a report by Nakamura et al. (Appl. Phys. Lett. **64** (1994) 1687) of high-luminosity blue LED, from GaN and related alloys. Both of these followed very shortly after the achievement of good p-type ZnSe by Park et al. (Appl. Phys. Lett. **57** (1990) 2127) and p-type GaN by Amano et al. (Jpn. J. Appl Phys. **28** (1989) L2112) and by Nakamura et al. (Jpn. J. Appl Phys. **31** (1992) L139; **31** (1992) 1258), where it had been very difficult, for both materials, to obtain p-type conductivity.

Since then, progress in the GaN area has been spectacular, with estimated sales in 2006 of $5 Billion. LEDs have been produced in blue, violet, and UV as well as in high-brightness, with particular emphasis on white (via phosphors). These have a long life-time. They have a myriad uses, including traffic lights, automobile lightning, back-lightning for mobile phones, flashlights, lighting of superstructure of bridges, outdoor displays, etc. Lasers are being used for improved optical storage density and resolution (e.g., for DVDs) as well as for the ability for chemical- and biohazard substance detection.

Nevertheless despite all the successes of GaN based materials (e.g., UV and violet laser diodes for 390–420 nm and efficient LEDs to 530 nm), there are remaining problems, which are difficult to solve. Fundamental aspects are relatively poor p-type doping and lack of good substrates. Consequently, these materials have not given adequate emission in the important pure green (560–565 nm) spectral region. It is in this region that the human eye has its maximum response, with obvious applications to displays. Other important applications are white emission without phosphors and plastic optical fiber networks. This is why II-VI ZnSe-based wide bandgap materials remain of high interest. Spectral response of these materials in the deep green critical spectral region is excellent, although they also have the problem of p-type doping. It worth noting that white LEDs without phosphors, based on ZnSe alloys, with lifetime of 10,000 hours have recently been reported (T. Nakamura, Electr. Eng. Japn. **154** (2006) 42). In addition, there is high present

Wide Bandgap Light Emitting Materials and Devices. Edited by G. F. Neumark, I. L. Kuskovsky, and H. Jiang
Copyright © 2007 WILEY-VCH Verlag GmbH & Co. KGaA, Weinheim
ISBN: 978-3-527-40331-8

interest in ZnO for light emitting applications because of its large exciton binding energy (60 meV), which results in very efficient emission near the band edge at room temperatures as well as its relatively lower index of refraction, which permits more efficient extraction of light from ZnO due to the large critical angle for total internal reflection. However, numerous challenges remain in utilizing ZnO in lightning applications with the principal challenge being obtaining efficient p-type doping.

The present book consists of two parts, Part 1 on GaN and related issues and Part 2 on wide bandgap II-VIs. Articles in Part 1 discuss overall progress in nitride light emitters (Chapter 1), GaN-based LEDs on novel substrates (Chapter 2), and miniature GaN lasers (Chapter 3). Part 2 is devoted to II-VI wide bandgap compounds, specifically, ZnSeTe alloys (Chapter 4) and ZnO (Chapter 5).

Chapter 1 summarizes recent progress in III-Nitrides light emitters with the emphasis on UV laser and LEDs, including those fabricated from nonpolar oriented materials. Chapter 2 covers key growth issues, design considerations, and the operation of III–nitride LEDs on sapphire and other substrates, including Si, SiC, bulk GaN and AlN. In Chapter 3 the recent progress in III-nitride micro-size structures and light emitters, which are important for future optical circuit elements, is summarized with emphasis on fabrication and optical properties. Some of the applications of micro-emitters for boosting output power of LEDs are also discussed. Chapter 4 devoted to the latest developments in optical properties of Zn–Se–Te grown by migration enhanced epitaxy with sub-monolayer quantities of Te with focuses on ZnTe/ZnSe type-II QDs, including the observation of the optical Aharonov–Bohm effect. Chapter 5 summarizes optical properties of ZnO and its alloys, including very recent results on novel ZnCdO alloys. Special attention is paid to rarely discussed issues such as the index of refraction, including the methods used to measure it.

The Editors, July 2007

Contributors

Dr. Xian-An Cao
One Research Circle
GE Global Research Center
KWC 1811
Niskayuna, NY 12309
USA

Dr. Yi Gu
Department of Materials Science
    and Engineering
Northwestern University
Evanston, IL 60208
USA

Prof. Dr. Hongxing Jiang
Department of Physics
Kansas State University
Cardwell Hall
Manhattan, KS 66506-2601
USA

Dr. Igor Kuskovsky
Department of Physics
Queens College of CUNY
65-30 Kissena Blvd.
Flushing, NY 11367
USA

Prof. Jingyu Lin
Department of Physics
Kansas State University
Cardwell Hall
Manhattan, KS 66506-2601
USA

Dr. John Muth
North Carolina State University
ECE Dept, MRC Bldg. RM 234 E
2410 Campus Shore Drive
Raleigh, NC 27695
USA

Prof. Dr. Gertrude F. Neumark
Department of Applied Physics and
    Applied Mathematics
Columbia University
500 W. 120th Street, Rm. 1137
New York, NY 10027
USA

Dr. Andrei Osinsky
SVT Association, Inc.
7620 Executive Drive
Eden Prairie, MN 55344
USA

Dr. Tao T. Wang
Department of Electronic and
    Electrical Engineering
University of Sheffield
Mappin Street
Sheffield, S1 3JD
UK

Wide Bandgap Light Emitting Materials and Devices. Edited by G. F. Neumark, I. L. Kuskovsky, and H. Jiang
Copyright © 2007 WILEY-VCH Verlag GmbH & Co. KGaA, Weinheim
ISBN: 978-3-527-40331-8

Part I

1
III–Nitride Light-Emitting Diodes on Novel Substrates

Xian-An Cao

1.1
Introduction

During the past decade, III–nitrides, which form continuous and direct bandgap semiconductor alloys, have undergone a phenomenal development effort, and have emerged as the leading materials for light-emitting diodes (LEDs) with peak emission spanning from green through blue to ultraviolet (UV) wavelengths [1, 2]. High-brightness (HB) green and blue LEDs along with AlInGaP red and yellow LEDs complete the primary color spectrum and enable fabrication of large-scale full-color displays. Near-UV and blue LEDs, when used in conjunction with multiband or yellow phosphors, can produce white light and are therefore very attractive for solid-state lighting applications [3]. There are also plentiful ongoing endeavors to push emission wavelengths into the deep-UV regime for numerous applications including bioaerosol sensing, air and water purification, and high-density data storage.

One of the most defining features of the nitride material system is the lack of high-quality bulk GaN or AlN substrates. To date, all commercially available III–nitride LEDs are grown heteroepitaxially on foreign substrates such as sapphire and SiC. Si has also received some attention as the substrate for low-power LEDs due to its clear advantages of low cost and high quality. Many efforts have been devoted to developing high-quality buffer layers to accommodate the mismatch in lattice constant and thermal expansion coefficient between the epilayers and substrates. The presence of a high density of threading dislocations and large residual strain in the heteroepitaxial structures, along with strong piezoelectricity and large compositional fluctuation of the nitride alloys, give rise to some unique electrical and optical characteristics of current III–nitride LEDs [4].

Bulk GaN and AlN would be a nearly perfect match to LED heterostructures, and meet most substrate requirements. Homoepitaxial growth significantly reduces defect density and strain, and offers better doping and impurity control. These incentives are the driving force behind recent progress toward producing bulk GaN and AlN crystals [5, 6]. Some free-standing GaN substrates are now

Wide Bandgap Light Emitting Materials and Devices. Edited by G. F. Neumark, I. L. Kuskovsky, and H. Jiang
Copyright © 2007 WILEY-VCH Verlag GmbH & Co. KGaA, Weinheim
ISBN: 978-3-527-40331-8

commercially available, and preliminary results of homoepitaxy are encouraging. However, before large-area low-cost GaN wafers become available, sapphire and other foreign substrates will remain the common substrates for nitride LEDs due to the well-established heteroepitaxy technology.

In this chapter, key growth issues, design considerations, and the operation of III–nitride green, blue and UV LEDs on sapphire are described. An overview is then given that describes the growth and performance of nitride LEDs on other novel substrates, including Si, SiC, and bulk GaN and AlN. The influence of the substrates on the microstructual properties, electrical characteristics, light emission, and light extraction of the LEDs is discussed.

1.2
LEDs on Sapphire Substrates

1.2.1
LED Heteroepitaxy

Most commercially available III–nitride LEDs are grown on the *c*-plane of sapphire substrates. Large-area and high-quality sapphire is widely available in large quantities, and is fairly inexpensive. To date, much more knowledge and experience have been accumulated in the technology of growing III–nitrides on sapphire than on other substrates. Another advantage of using sapphire is its transparency to UV and visible light, reducing the parasitic light loss in the substrate. However, there are a few disadvantages associated with sapphire substrates. First, a large mismatch in lattice constant (~15%) and thermal expansion coefficient between nitride materials and sapphire gives rise to a high density of threading dislocations and biaxial stress in epitaxial layers. Second, sapphire is electrically insulating, making it necessary to fabricate LEDs in a lateral configuration. In this case, current spreading is a key device design consideration (described in Section 1.2.2). Third, sapphire has a relatively poor thermal conductivity, limiting heat dissipation in top-emitting LEDs. This, however, is less of a problem in flipchip LEDs, where heat is removed through the p-type contact.

Metal–organic chemical vapor deposition (MOCVD) has evolved as the dominant technique for growing III–nitride LEDs, not only on sapphire but also on other substrates [7]. However, the design of MOCVD reactors for nitrides is much less mature than for conventional III–V semiconductors. Currently both commercial and home-built reactors are used, and the growth process conditions vary widely. MOCVD is a nonequilibrium chemical process, in which gaseous precursors are injected from a precise gas-mixing manifold into a cold-wall reactor, where they react on a heated substrate. LED epitaxy is usually conducted under a low pressure. The common precursors include trimethylaluminum (TMAl), trimethylgallium (TMGa), and trimethylindium (TMIn) as the metal sources, and ammonia as the N source. Silane and bis-cyclopentadienyl-magnesium (Cp_2Mg) are used for n- and p-type doping, respectively. Hydrogen or nitrogen is used as the carrier gas.

The ideal growth temperatures for different layers of the LED structure are different: GaN is grown at 1000–1100 °C, AlGaN requires a slightly higher temperature, and InGaN is grown at a much lower temperature ~700–800 °C.

The realization of HB III–nitride LEDs on sapphire is based upon two epitaxy technology breakthroughs. The first was the demonstration of p-type conductivity in GaN. As-grown Mg-doped GaN has a very high resistivity due to the formation of Mg–H complexes. It was found that Mg can be activated by dissociating H from Mg using low-energy electron-beam irradiation [8] or annealing at >750 °C [9]. The second breakthrough was growing high-quality nitrides on sapphire using a thin low-temperature buffer layer [10]. The crystalline quality of GaN films grown directly on sapphire is generally poor due to the large mismatch between GaN and sapphire. Early work by several groups showed that GaN layers grown atop a thin AlN or GaN buffer layer greatly improved surface morphologies and crystalline quality [10, 11]. A buffer layer with a thickness of <100 nm, usually grown at ~500 °C, is critical to defect reduction and subsequent two-dimensional (2D) growth of device structures. Prior to the growth of the buffer, the sapphire substrate is usually nitridated by exposure to ammonia gas in the reactor [12, 13]. The nitridation process promotes GaN and AlN nucleation on sapphire and further improves the quality of the overlayer.

The first blue LED, reported by Nichia, consisted of InGaN/AlGaN double heterostructures (DHs) [14]. The active region was an InGaN layer codoped with Si and Zn, and an impurity-related transition was responsible for the blue emission. Their second-generation blue LEDs had an undoped InGaN single-quantum-well (SQW) active region, which exhibited efficient direct bandgap emission [15]. Figure 1.1(a) shows a typical layer structure of state-of-the-art green and blue LEDs, which consists of an InGaN/GaN multiple-quantum-well (MQW) active region [16, 17], an n-GaN lower cladding layer, a p-AlGaN upper cladding layer, and a p$^+$-GaN top contact layer. The p-AlGaN cladding layer is necessary to prevent electrons from escaping from the active region. The top surface is usually doped heavily with Mg to reduce the p-contact resistance. Similarly, deep-UV LEDs would have an AlGaN MQW active region sandwiched between n- and p-type AlGaN cladding layers with

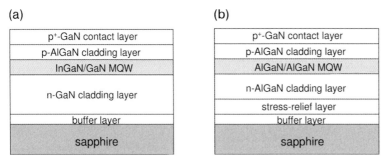

Fig. 1.1 Schematic of typical layer structures of state-of-the-art
(a) InGaN-based visible LEDs and (b) AlGaN-based UV LEDs
on sapphire substrates.

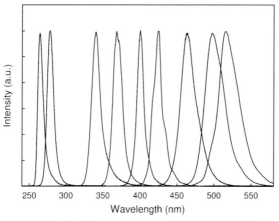

Fig. 1.2 Electroluminescence spectra of III–nitride MQW LEDs grown on sapphire substrates using MOCVD.

a higher Al content (Fig. 1.1(b)) [18, 19]. Figure 1.2 illustrates a series of electroluminescence (EL) spectra of III–nitride MQW LEDs made on sapphire with the MOCVD technique and with peak emission ranging from green to deep-UV. The green and blue LEDs have a much larger full width at half-maximum (FWHM) of ~30 nm than the UV LEDs. The spectral broadening is believed to arise from large compositional inhomogeneity in the InGaN active regions [4].

During the LED overgrowth, the epilayer is essentially relaxed rather than being strained to lattice-match to the sapphire. However, a large biaxial compressive stress may be generated upon cooling to room temperature due to the larger thermal expansion coefficient of sapphire [20]. The actual magnitude and sign of the stress are a function of the growth conditions, and depend largely on the thickness and doping level of the thick n-type cladding layer. It has been found that excessive Si doping may change the stress from compressive to tensile [21], which promotes wafer bowing and film cracking, and limits the maximum size of wafers and the thickness of LED structures. Compared to GaN epilayers, AlGaN films grown on sapphire are more subject to a large residual stress. To improve strain management, state-of-the-art deep-UV LEDs are grown on a thick AlN template with an additional AlN/AlGaN superlattice stress-relief layer [18, 19]. Zhang et al. [22] reported a pulsed atomic-layer epitaxy technique, which considerably suppresses gas-phase reaction, enhances adatom surface migration, and produces AlN and AlGaN template layers with reduced alloy disorder and improved surface morphology. The high-quality stress-relief layer has proven to be crucial for the subsequent growth of High-Al content AlGaN LED structures.

Despite the use of a buffer layer to accommodate the large lattice mismatch between nitrides and sapphire, a very high level of threading dislocations (10^8– $10^{10}\,cm^{-2}$) is present within LED heterostructures [23]. Some traverse vertically from the epilayer/substrate interface to the top layer and, depending on the growth conditions of the capping layer, may terminate by forming various types of surface

 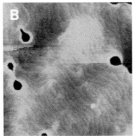 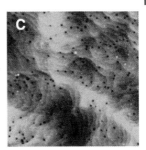

Fig. 1.3 AFM images (2 μm × 2 μm) of three representative InGaN/GaN MQW LEDs grown on sapphire using MOCVD.

defects. Figure 1.3 shows atomic force microscopy (AFM) images of three InGaN/ GaN MQW LEDs on sapphire with different surface morphologies. The root-mean-square (rms) surface roughness of these samples is in the range of 0.4– 0.8 nm over a $2 \times 2 \mu m^2$ area. The surface of LED A is microscopically rough but free of obvious pits, whereas LEDs B and C present a large number of surface defects. On sample B, there are $\sim 1.5 \times 10^8 cm^{-2}$ pits with hexahedral cone morphology and a size ~ 100 nm. LED C shows a swirled step structure and $4 \times 10^9 cm^{-2}$ small surface pits, which are caused by the intersection of the top surface with the dislocations [24]. Cross-sectional transmission electron microscopy (TEM) showed that the densities of dislocations reaching the MQW active region in LEDs A, B and C were $6 \times 10^8 cm^{-2}$, $3 \times 10^9 cm^{-2}$, and $5 \times 10^9 cm^{-2}$, respectively. In LED A, dislocation bending was found at 200–350 nm after the buffer layer, and a relatively small number of dislocations, mainly of edge character, propagated to the top layer. The hexagonal pits in LED B, usually called V-defects, were found to form in either the p-GaN capping layer or the MQW region, and were connected to threading dislocations. Large strain at the GaN/InGaN interfaces and In-rich regions [25, 26], or impurity complexes, are believed to be the cause of their formation [27].

The dislocation densities in III–nitride LEDs grown on sapphire are far higher than those observed in working LEDs based on conventional III–V semiconductors. GaAs-based LEDs with a dislocation density $>10^4 cm^2$ would not show any band-edge emission [28]. The fact that efficient blue and green LEDs can be made with highly defective InGaN materials suggests that threading dislocations do not act as efficient nonradiative recombination centers [29, 30]. This is supported by the finding that blue and green LEDs grown on a high-quaity, laterally overgrown GaN template with a dislocation density of $\sim 7 \times 10^6 cm^{-2}$ had an external quantum efficiency similar to LEDs grown on a regular buffer layer [31]. It is now well accepted that In-rich quantum-dot-like (QD-like) regions self-formed in InGaN alloys due to strong compositional fluctuation, enhance carrier localization and radiative processes [4]. The localization effects, which are however lacking in high-quality AlGaN alloys, result in some unique EL behaviors of InGaN LEDs, and will be detailed in Sections 1.2.4 and 1.2.5.

1.2.2
Current Spreading

Due to the insulating substrate, InGaN/GaN LEDs grown on sapphire must be fabricated in a lateral configuration. A mesa is defined using plasma etching so that the n-type electrode can be deposited on the exposed n-GaN cladding layer, whereas the p-type contact is formed atop the p-GaN layer. The resistivity of the top p-GaN layer is typically several orders of magnitude higher than that of the n-type cladding layer. It is therefore necessary to add an additional conducting layer to spread current to regions not covered by the p-type bonding pad. In top-emitting LEDs, current spreading on the p-side usually relies on the use of a semitransparent contact covering the entire p-GaN [32, 33]. The current spreading layer also functions as an ohmic contact and light extraction window, and therefore must be transparent to the emitted light.

To reduce current nonuniformity, lateral current paths as determined by the spacing between the p-type and n-type electrodes should be smaller than the current spreading length, which is the length where current density drops to 1/e of that under the p-pad or at the mesa edge. The current spreading length L_s in a top-emitting LED is given by

$$L_s = \left(r_c + \rho_p t_p\right)^{1/2} \left|\frac{\rho_n}{t_n} - \frac{\rho_t}{t_t}\right|^{-1/2} \tag{1.1}$$

where ρ_p, ρ_n, ρ_t, t_p, t_n, and t_t are the respective resistivity and thickness of the p-GaN, n-GaN, and semitransparent contact, and r_c is the specific contact resistance of the p-contact [33]. It is clear that a uniform current distribution can be achieved when the n-type and p-type current spreaders have an identical sheet resistance (i.e., $\rho_t/t_t = \rho_n/t_n$) [32]. Current tends to crowd toward the p-pad or mesa edge adjoining the n-type electrode when this condition is not met. Current crowding may lead to a nonuniform light emission and self-heating, thus reducing the quantum efficiency and accelerating LED degradation. With increasing mesa size, current crowding becomes more severe. Novel mesa geometries, such as interdigitated mesas or multiple isolated mesas with a width less than L_s, must be used to alleviate this problem [34]. These types of designs also enjoy the advantage of good scalability, which is critical for developing large-area high-power LEDs.

A number of semitransparent contacts comprising a thin metal film have been investigated [35–39], among which a bilayer Ni/Au thin film is the most extensively used. It was found that the contact resistance of an Ni/Au contact to p-GaN can be substantially reduced by annealing the contact in an oxygen ambient to form NiO/Au [35]. The resultant NiO embedded with Au islands is believed to be a low-barrier contact to p-GaN [36], with a specific contact resistance in the 10^{-3}–$10^{-4}\,\Omega\,cm^2$ range. The oxidized Ni/Au is electrically conducive and semitransparent at visible wavelengths. Both the conductivity and transparency are strong functions of the Ni/Au content ratio. Figure 1.4 shows the sheet resistance and light transmission of Ni (5 nm)/Au films with varying Au thickness before and

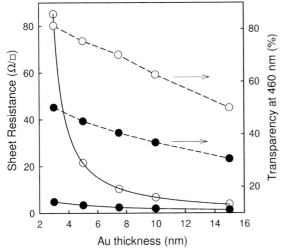

Fig. 1.4 Sheet resistance (solid lines) and light transmission (dashed lines) of Ni(5 nm)/Au films with varying Au thickness before (solid dots) and after (open dots) 550 °C annealing in air.

after a 550 °C anneal in air. The transparency is improved by ~60% after the anneal, and decreases rapidly with increasing Au thickness, whereas the resistance increases dramatically with decreasing Au content. Providing that the sheet resistance of the n-GaN layer in typical blue LEDs is ~20 Ω/□, the optimal Au thickness is 5–6 nm from the current spreading viewpoint. At this thickness, the Ni/Au film is >70% transparent at 460 nm, and forms a low-resistance ohmic contact to p-GaN.

In flipchip LEDs, the current on the p-side spreads in a thick ohmic metal, which also functions as a reflective mirror. Assuming negligible resistance of the p-metal, the current spreading length can be expressed as [40]:

$$L_s = \left((r_c + \rho_p t_p) \frac{t_n}{\rho_n} \right)^{1/2} \tag{1.2}$$

In this case, current tends to crowd at the edge of the mesa contact, and current density decreases exponentially with increasing distance from the mesa edge. Equation (1.2) shows that L_s can be increased by reducing the resistivity or increasing the thickness of the n-type cladding layer. In AlGaN-based deep-UV LEDs, L_s may be one order of magnitude smaller than in typical blue LEDs due to the low conductivity of high-Al AlGaN materials. Flipchip UV LEDs grown on sapphire are therefore more subject to current crowding and localized heating problems [41]. Interdigitated mesa structures or small mesa arrays must be employed to mitigate these problems.

1.2.3
Carrier Transport

Transporting electrons and holes from the current spreading layers to the QW active region is an essential step to generate light in the active region. Interfacial potential barriers and defect states in the cladding layers are expected to have a pronounced influence on the carrier transport dynamics. In a semiconductor p–n junction, the forward I–V characteristics at moderate bias can be described by the conventional Shockley model:

$$I = I_0 \exp(qV/nkT) \tag{1.3}$$

where I_0 is the saturation current, q is the electron charge, k is Boltzmann's constant, and T is the absolute temperature [42]. The ideality factor n, which is correlated with the slope of the semilogarithmic I–V plot, has a value between 1 and 2, indicating the coexistence of diffusion current ($n = 1$) and recombination current through bandgap states in the space-charge region ($n = 2$). In LEDs with an MQW active region, the ideality factor value associated with the diffusion process may be different. Carriers must diffuse over about one-half of the total barrier to be injected into the QWs where they recombine radiatively or nonradiatively. Taking this into consideration, an ideality factor close to 2 can be expected in high-quality InGaN/GaN QW LEDs [43].

In addition to thermal diffusion and recombination, tunneling may be another important mechanism of carrier transport in typical nitride LEDs. This is the case for two reasons. First, a high density of bandgap states is present in the space-charge region providing the tunneling paths. Second, the n- and p-type cladding layers are usually heavily doped, and the quantum barrier layers may also be doped with Si to screen the large piezoelectric field. The depletion layer at the junction is therefore relatively thin, enhancing carrier tunneling through the potential barriers at heterointerfaces. Following the analysis of Dumin et al. [44], the nonradiative tunneling current in a p–n junction involving single level traps depends exponentially on the applied voltage:

$$I = CN_t \exp(qV/E) \tag{1.4}$$

where C is a constant containing the built-in potential, N_t is the trap density, and the energy parameter E in a p^+–n junction is given by:

$$E = \frac{2hq\sqrt{N_D}}{\pi^2 \sqrt{m^* \varepsilon_s}} \tag{1.5}$$

where m^* is the effective carrier mass, ε_s is the semiconductor dielectric constant, N_D is the donor concentration, and h is Planck's constant. One simple model that can be considered is that of electron tunneling to traps in the p-type cladding layer, sometimes evidenced by the associated defect radiative emission, or hole tunneling to bandgap states in the MQW region, followed by radiative or nonradiative recombination. The excess current could also result from a more complex process involving multiple-level states. This is especially possible in III–nitride LEDs because the defect density is sufficiently high. Possible sources of electrically active states include dislocations, impurities, Ga or N vacancies, and antisites. As seen in Eq. (1.4), the tunneling current has little temperature sensitivity. The slope of the semilogarithmic *I–V* plot is expected to be temperature-independent, which is in contrast to temperature-dependent diffusion and recombination currents.

Figure 1.5(a) compares the forward *I–V* characteristics of the LEDs whose micro-structural properties were described in Section 1.2.1. At high injection currents ($>10^{-3}$ A), the details of carrier transport cannot be identified due to a high series resistance. At low and moderate bias, the *I–V* behaviors of LEDs B and C can be represented by Eq. (1.4), suggesting the dominance of tunneling current. Both LEDs show two main exponential segments with different slopes, which may result from tunneling processes involving different deep-level states. The energy parameter *E* has a room-temperature value of ~220 meV at voltages of 0–1.7 V and 72–105 meV at voltages of 1.8–2.6 V. At moderate bias, the unrealistic ideality factors are 2.8 and 4 for LEDs B and C, respectively.

The low-bias forward *I–V* characteristic of LED A is also dominated by tunneling current. However, at moderate bias above 2.1 V, it can be modeled by the conventional drift-diffusion model as $I = I_0 \exp(eV/1.6kT)$. The dominance of diffusion–recombination current reflects the high quality of LED A, in agreement with earlier structural analysis. Defect-assisted tunneling current is significantly suppressed. This current component has a slope proportional to $1/1.6kT$, as seen in Fig. 1.5(b). Also shown in this figure are the changes of the characteristic energy at moderate bias in LEDs B and C with increasing temperature. While the *E* value in LED C is mostly temperature-independent, it decreases slightly with increasing tempera-

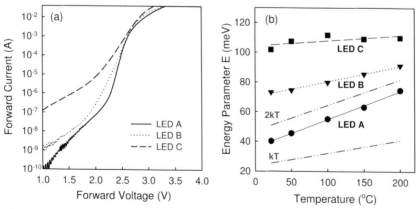

Fig. 1.5 (a) Forward *I–V* characteristics of three InGaN/GaN LEDs measured at room temperature. (b) The characteristic energy *E* at moderate bias as a function of measurement temperature.

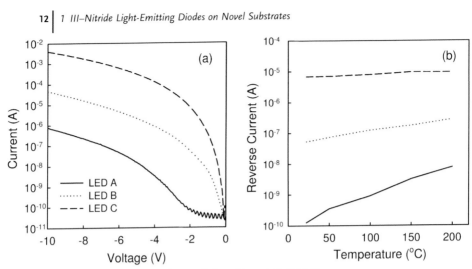

Fig. 1.6 (a) Reverse *I–V* characteristics of three InGaN/GaN LEDs measured at room temperature. (b) Reverse current at − 2 V as a function of measurement temperature.

ture in LED B, indicating an increased contribution from diffusion–recombination current. The carrier injection is therefore a combination of tunneling, diffusion, and recombination processes. These results confirm a strong correlation between microstructural quality and the mechanism of current transport in III–nitride LEDs.

Reverse dark current in III–nitride LEDs usually scales with the junction area due to the dominant bulk leakage [45], and is many orders of magnitude higher than classical diffusion and generation–recombination currents in wide bandgap semiconductors. Figure 1.6(a) shows the reverse *I–V* characteristics for the three LEDs measured at 25 °C. The currents in LEDs B and C are strongly voltage-dependent but temperature-insensitive, as seen in Fig. 1.6(b). These are characteristic features of defect-assisted tunneling. Similar behaviors have also been found in conventional III–V diodes [46, 47], GaN p–n diodes [48], and double heterostructure (DH) blue LEDs [49]. At high reverse voltage, the data are in good agreement with band-to-band tunneling as predicated by the Zener tunneling model [46]. The electrons may tunnel from the p-GaN valence band to the n-GaN conduction band through to a thin depletion layer. The leakage current in LED A is several orders of magnitude lower. Carrier tunneling is not dominant until reverse bias >2 V, where there is a sudden increase in the slope. At bias <2 V, the leakage current is roughly an exponential function of temperature, indicative of the presence of thermal currents such as space-charge generation current.

1.2.4
Carrier Confinement and Localization

The confinement of injected carriers inside the MQW region increases the overlap of electrons and holes, resulting in much faster and more efficient radiative recombination. To reduce the nonradiative recombination rate, the thickness of InGaN QWs should be less than the minority carrier diffusion length and is typically a few nanometers. The thin QWs also give rise to a high carrier density, which is preferred to suppress the nonradiative process by saturating defect states in the active region. The active layer is also designed to be thin to improve light extraction by reducing the internal band-to-band absorption.

In contrast to light-emitting devices based on conventional III–V semiconductors, InGaN-based LEDs are surprisingly efficient despite the existence of a high density of microstructural defects [29]. It is generally accepted that carrier localization effects that arise from spatially inhomogeneous indium distribution and QW thickness fluctuation play an important role in spontaneous emission from InGaN QW structures [4, 30, 50–55]. The InGaN compositional disorder occurs due to large differences in thermal stability and lattice parameters between GaN and InN. The localization effects improve the radiative recombination in two ways. First, QD-like In-rich regions in InGaN alloys trap carriers, forming localized excitons. This further enhances the overlap of electrons and holes, and their recombination rates. Second, localization effects prevent carriers from reaching defects, and thus reduce nonradiative recombination. Nanoscale QD-like structures in InGaN alloys have been measured directly by high-resolution TEM [53]. Submicron emission fluctuation resulting from In segregation was clearly observed from cathodoluminescence (CL) images [54]. The mean size of the QDs increases with increasing In content [55], indicating larger In compositional fluctuation and stronger localization effects in higher In content InGaN materials.

One characteristic feature of InGaN-based LEDs is a pronounced blueshift of the EL peak with increasing drive current. This can be attributed to the band filling of localized states at potential minima in the QW plane [4]. Figure 1.7 shows the EL spectra of an InGaN MQW green LED and an AlGaN MQW UV LED at 300 K as functions of injection current. The spectra of the green LED are much broader due to strong alloy broadening, and show a large blueshift of 54 meV as the current is increased from 0.1 mA to 20 mA. In contrast, the emission peak of the UV LED has a small FWHM of ~10 nm and is almost independent of injection current. These features suggest much smaller compositional fluctuation and weaker localization effects in the AlGaN MQWs. Carriers are, therefore, more likely to be trapped and recombine nonradiatively at defect states. This may partly explain why state-of-the-art deep-UV LEDs are much less efficient and less reliable than commercial InGaN LEDs. It has been found that the optical properties of AlGaN materials can be markedly improved by introducing a small amount of In to form quaternary AlInGaN alloys with enhanced localization effects [56–59]. In addition, by varying the compositions of quaternary alloys, the bandgap and lattice constant can be adjusted independently to achieve lattice-matched structures with a reduced dislocation density and piezoelectric field [60]. Efficient UV LEDs with peak wavelength 305–350 nm have been demonstrated based on AlInGaN MQW structures [57–59].

Another factor that may account for the EL blueshift is a large static electric field (up to a few MV cm^{-1}) in the InGaN QWs due to spontaneous and piezoelectric

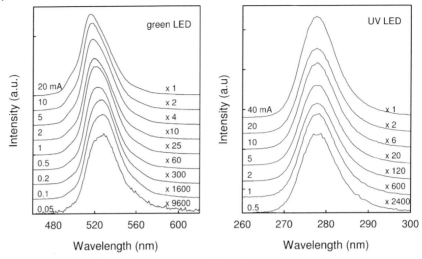

Fig. 1.7 EL spectra of an InGaN-based green LED and an AlGaN-based UV LED at 300 K as a function of injection current. The spectra are shifted in the vertical direction for clarity.

polarization [61, 62]. The latter arises from biaxial strain remaining in the heterostructures due to the lattice mismatch between the well and barrier layers. The electric field tilts the energy bands and separates electron–hole pairs into triangular potential wells formed at opposite sides of the QWs. When carriers are injected into the QWs and screen the polarization charges, the overlap of the electron–hole wavefunction is recovered, resulting in a blueshift of emission peak [63, 64]. In LEDs with thin InGaN QW active layers (≤3 nm), efficient carrier confinement and localization partially overcome the disadvantage of the piezoelectric effect. It has been found that doping the GaN quantum barrier layers with Si can effectively screen polarization-induced charges at the heterointerfaces [62, 65]. This technique is now extensively employed in commercial blue and green LEDs [66]. The EL blueshift in the green LED, as seen in Fig. 1.7, is therefore due largely to the band filling of localized states.

A better understanding of the role of localized states in the radiative process in InGaN-based LEDs can be gained by means of temperature-dependent luminescence measurement. Three LEDs (green, blue and UV) with a similar epitaxial structure, but varying In and Al contents in the MQWs and cladding layers, were characterized in the temperature range of 5–300 K. The respective nominal In mole fractions in the active regions were 0.35, 0.20 and 0.04. Figure 1.8 shows temperature-induced shifts of the peak energy of the LEDs operating at 1 mA. With decreasing temperature, a blueshift and then a redshift is seen for all LEDs, although the amounts of the shifts are different. The redshift in the UV LED between 77 and 200 K is as large as 50 meV. Considering that the temperature-

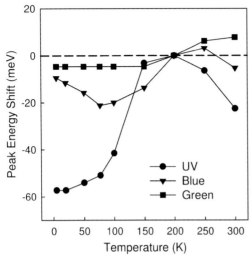

Fig. 1.8 Peak energy shift in InGaN/GaN MQW LEDs as a function of measurement temperature.

induced change of the GaN bandgap energy is ~26 meV, the actual displacement of the peak energy relative to the band edge in this LED is −76 meV. The unusual energy redshift can be explained by carrier relaxation [67]. As the temperature is decreased, the carrier lifetime increases due to reduced nonradiative rates, allowing more opportunity for carriers to relax into lower-lying localized states. The energy redshift, and therefore the degree of carrier relaxation, decreases as the In content in the active region increases. In the green LED, the peak redshifts by only ~10 meV in the temperature range of 150–250 K and stabilizes at 2.37 eV at lower temperatures. This behavior can be interpreted as evidence of strong localization effects in this LED. A large number of localized states in the MQWs are responsible for efficient carrier capture and recombination over the entire temperature range. In contrast, the UV LED demonstrates different emission mechanisms within different temperature ranges. While localized state recombination appears to be important below 150 K when carriers are transferred to lower-energy states, band-to-band transition becomes dominant at higher temperatures as the localized carriers are thermalized. Indeed, with increasing temperature above 200 K, the EL peak of the UV LED exhibits a redshift, which follows the characteristic temperature dependence of the GaN bandgap shrinkage [68]. The thermal energy at 150 K ~ 13 meV can be estimated as the magnitude of the localization energy in the UV LED.

Figure 1.9 shows the L–I characteristics of the LEDs measured at 5 K. Light intensity L is linear with current I in the low injection regime for all devices, indicating dominant radiative recombination and a constant quantum efficiency. As current increases, the light output shows a trend of saturation. The dependence of light output on forward current becomes sublinear in the high injection regime,

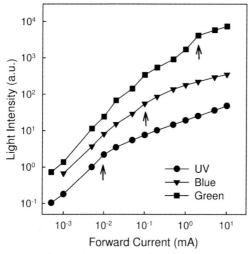

Fig. 1.9 The *L–I* characteristics of InGaN/GaN MQW LEDs at 5 K. The arrows indicate the current levels where the light output tends to saturate.

following $L \sim I^m$ ($m \sim 0.4$). The transition point where the EL intensity tends to saturate moves to higher injection currents as the In content increases. This is evidence that carrier capture becomes inefficient at low temperatures. In the UV LED, there are only a limited number of localized states, and therefore only a small portion of the injected carriers can transfer to localized states, whereas a significant number of carriers may escape from the InGaN QWs. As expected, there are a large number of localized states in the green LED, offering carriers more opportunities to be captured. The *L–I* data can be fitted to the solution of the rate equation, which describes carrier capture and decapture processes in the MQWs at steady state:

$$\frac{dN}{dt} = (N_0 - N)j_c\sigma_{cap} - N\left(\frac{1}{\tau_{decap}} + \frac{1}{\tau_{rad}} + \frac{1}{\tau_{nonrad}}\right) = 0 \qquad (1.6)$$

where N_0 is the total number of localized states, N is the number of captured carriers, j_c is the flux of carrier injection, σ_{cap} is the effective cross-section for carrier capture, and τ_{decap}, τ_{rad} and τ_{nonrad} represent the lifetimes associated with carrier decapture, radiative and nonradiative recombination, respectively. Considering that the nonradiative lifetime of localized carriers τ_{nonrad} and the carrier decapture time τ_{decap} at low temperatures (<150 K) are quite long, and that the EL intensity L is proportional to N/τ_{rad}, we obtain

$$L = \frac{N_0}{(1/j_c\sigma_{cap}) + \tau_{rad}} \qquad (1.7)$$

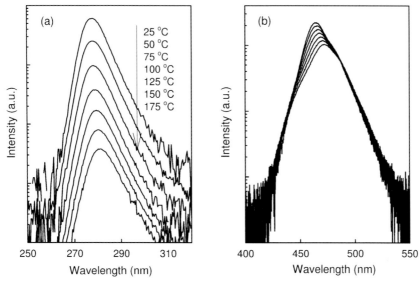

Fig. 1.10 Temperature-dependent EL spectra of (a) a deep-UV LED and (b) a blue LED at 10 mA.

From Eq. (1.7), L is given by $L = N_0 j_c \sigma_{cap}$ in the limit of $j_c \to 0$. On the other hand, L saturates to the value of N_0/τ_{rad} at high currents ($j_c \to \infty$). The radiative decay time is expected to be longer in the green LED due to the larger size of QD structures [69]. The output power of the green LED at high currents (>10 mA) is ~200 times higher than that of the UV LED, as shown in Fig. 1.9, suggesting that the density of the related localized states in the green LED is more than two orders of magnitude higher.

The localization effects also have a significant impact on the LED performance at elevated temperatures. Figure 1.10 illustrates the evolution of the EL spectra of a deep-UV (280 nm) LED and blue (465 nm) LED with increasing temperature. As temperature is increased from 25 °C to 175 °C, the emission of the UV LED shows a much sharper decrease, by a factor of 48, compared to only a 42% decrease for the blue LED. While the high-energy side of the blue LED spectra shows a slow decrease in emission intensity, the intensity and energy of the low-energy band, especially the tail, are almost temperature-independent. The low-energy emission is dominated by emission at localized states and less sensitive to the change in temperature. The localization effects remain strong even at 175 °C, as also revealed by a blueshift of the EL peak with increasing current, suggesting a localization energy >39 meV. A previous study using time-resolved photoluminescence (TRPL) showed that the localization energy in an InGaN material with a composition similar to the blue LED QWs was ~60 meV [70], corresponding to an indium compositional fluctuation of ~0.03. As a contrast, the low-energy and high-energy sides of the UV emission spectra exhibit similar temperature dependence of emission intensity and energy. The redshift of the UV peak follows the temperature dependence of the AlGaN bandgap as described by Varshni's equation

$$E_g(T) = E_g(0\,\mathrm{K}) + \alpha T^2/(T - \beta) \tag{1.8}$$

with values $\alpha = 5.08 \times 10^{-4}\,\mathrm{eV\,K^{-1}}$ and $\beta = 996\,\mathrm{K}$ [71]. These behaviors again suggest minimal localization effects in AlGaN alloys and that the UV emission is dominated by band-to-band transition. The much sharper decrease in EL intensity of the UV LED compared to the blue LED is due largely to its shallower AlGaN QWs and a higher defect density in the active region. The band offset between the well/barrier layers is ~0.28 eV in the UV LED, much smaller than ~0.71 eV in the blue LED. The above results also suggest that the lack of localization effects in AlGaN is another causal factor in the poor thermal performance of the UV LED, and that increasing carrier confining potentials will provide a critical means to improve UV LED performance.

1.2.5
Radiative and Nonradiative Recombination

The MQW structure greatly enhances carrier confinement and radiative recombination rates. However, nonradiative recombination can never be totally eliminated. Injected carriers may recombine nonradiatively through defect states in the QWs or escape from the QWs via tunneling or thermal motion and then recombine through defect states in the barrier and cladding layers. The unique material properties of InGaN, including strong localization effects and a large exciton binding energy, distinguish the radiative and nonradiative processes from other III–V materials. In high-quality blue and green LEDs, most carriers are captured by localized states in the QWs, leading to a dominant radiative process with an efficiency as high as 70–80%. With decreasing In content in the active region, the density of localized states decreases, and carriers have more chance to be captured by defect states. It has been demonstrated that the quantum efficiency of nitride LEDs decreases dramatically when the emission wavelength is shorter than 380 nm [72]. Particularly in AlGaN-based deep-UV LEDs, the radiative and nonradiative decay times are strongly affected by the presence of microstructural defects and polarization charge-induced field. The highest external quantum efficiency of 280 nm LEDs reported to date is less than 2% [2].

While most recombination processes via defect states are nonradiative, some deep-level transitions are radiative. Figure 1.11 shows current-dependent EL spectra of a blue LED whose *I–V* characteristics are dominated by carrier tunneling [see LED B in Fig. 1.5(a)]. At currents less than 0.02 mA (forward voltage <2.5 V), a broad deep-level emission centered at ~550 nm dominates. With increasing current, the QW emission at 460 nm increases rapidly, whereas the deep-level emission tends to saturate, suggesting that the yellow defect luminescence and carrier tunneling may be associated with the same prominent defect states in the

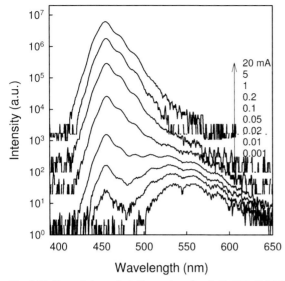

Fig. 1.11 Current-dependent EL spectra of an InGaN/GaN MQW blue LED.

space-charge region. Much stronger defect emission is usually observed in AlGaN UV LEDs [18, 73–75], where the defect state density is higher. Zhang et al. [76] reported a 280 nm LED with a defect emission band at ~320 nm, which was attributed to transitions involving deep acceptor levels in the p-type cladding layer. By increasing the carrier confinement potential, the ratio of QW and defect emission intensities at 20 mA was improved from 1.5 : 1 to 47 : 1.

The *L–I* characteristics of a LED in a specific injection regime can be fitted with a power law, $L \propto I^m$. The parameter *m* reflects the influence of defect states on carrier recombination processes [77]. As nonradiative recombination dominates, *L* shows a superlinear dependence on *I* ($m > 1$). In this case, the internal quantum efficiency increases with increasing current. When defect states are saturated and radiative recombination becomes dominant, a linear increase of *L* with *I* ($m \sim 1$) can be expected. The quantum efficiency becomes a constant. Figure 1.12 shows the *L–I* characteristics of an InGaN/GaN MQW UV LED measured at 5 K, 77 K, 150 K, and 300 K. The light output varies as I^2 at 300 K, indicating a significant influence of defects even at high current densities. At 150 K, even though some defects are frozen out, the impact of the defect states is still clearly seen at low currents (less than 0.1 mA). As temperature is further decreased, the transition point from nonradiative recombination to radiative recombination moves to lower injection currents. In high-quality blue and green LEDs, a linear relationship between *L* and *I* can be obtained in all injection regimes at room temperature [78], implying a much smaller role of defect states.

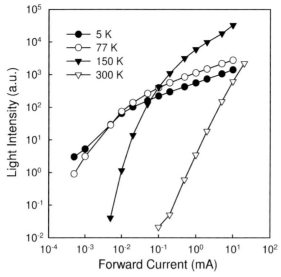

Fig. 1.12 The *L–I* characteristics of an InGaN/GaN MQW UV LED measured at 5 K, 77 K, 150 K, and 300 K.

1.2.6
Light Extraction

For LEDs having a quantum efficiency of unity, the recombination of every electron–hole pair generates one photon in the active region. The photons must escape from the semiconductor material in order to contribute to the total light emission. Just as with conventional III–V LEDs, light trapping inside LED chips is a primary limitation for producing efficient III–nitride LEDs. The large contrast in index of refraction between nitrides and the surrounding media gives rise to a small critical angle for total internal reflection as determined by Snell's law, and small light escape cones. A large portion of the generated light is trapped and rattles around until it is absorbed by the semiconductor, substrate or contact metals. The refractive index of GaN is ~2.5 at blue wavelengths, much larger than that of sapphire ($n = 1.78$) and a typical epoxy ($n = 1.5$). For LEDs grown on sapphire, light outside the escape cones is trapped within the thin epilayer. Based on ray-tracing simulation, we have found that the extraction efficiency of a planar blue LED on sapphire is about 12% without encapsulation, and about 27% with epoxy encapsulation.

In top-emitting LEDs, a significant amount of light is absorbed by the bondpads, semitransparent p-contact, and die-attach material. This type of loss can be prevented by employing a reflective p-type metal contact and a flipchip bonding scheme [79, 80]. In this case, the LED is flipchip mounted either on a submount or directly on a printed circuit board using an eutectic melt process, as illustrated in Fig. 1.13(a). All light is extracted through the transparent sapphire substrate. The flipchip configuration thus has a higher light extraction efficiency and is par-

(a) 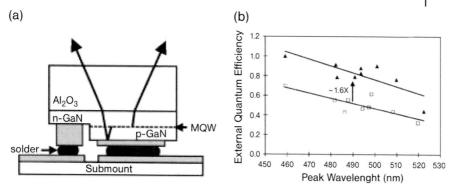 (b)

Fig. 1.13 (a) Schematic cross-section of an InGaN/GaN flipchip LED mounted on a submount. (b) Normalized external quantum efficiencies of top-emitting LEDs (open squares) and flipchip LEDs (solid triangles) tested at 350 mA pulse current. After Ref. [79].

ticularly important for developing high-efficiency UV (<370 nm) LEDs because the p-GaN contact layer and p-metal are strongly absorbing in the UV regime. Another advantage of flipchip LEDs over top-emitting LEDs is their superior thermal management. Heat generated in the junction can be removed efficiently through the thick p-metal and solder bumps, overcoming the disadvantage of the poor thermal conductivity of sapphire. Flipchip LEDs therefore suffer less from joule heating and can be driven at much higher currents.

The p-type metallization of flipchip LEDs has three functions: an ohmic contact, a current spreading layer, and an optical reflector. Silver is the metal of choice due to its high reflectivity at wavelengths longer than 400 nm, and its ability to form low-resistance ohmic contacts to p-GaN [80]. The whole metallization typically consists of a multilayer metal stack, including a diffusion barrier layer which prevents Ag from electromigration, and a solder layer for subsequent die attachment. Aluminum has a reflectivity greater than 90% at visible wavelengths, and thus can be utilized to form reflective n-type contacts. Figure 1.13(b) shows a plot of normalized external quantum efficiencies of top-emitting LEDs and flipchip LEDs tested at a 350 mA pulse current [79]. The LEDs with the same emission wavelengths were fabricated from the same epitaxial wafer. The flipchip LEDs have an external quantum efficiency 1.6 times higher than the top-emitting LEDs, suggesting a 60% increase in light extraction efficiency via flipchip packaging.

Many approaches have been explored to improve light extraction from LEDs on sapphire, including roughening chip surfaces [81, 82], texturing sapphire substrates [83, 84], integrating LEDs with a sapphire microlens [85] or photonic crystals [86, 87], and utilizing surface plasmon resonance [88]. The philosophy common to the first three approaches is to deliberately redirect the waveguided light, reduce total internal reflection, and consequently induce light emission in off-plane directions. Providing that the majority of the light is trapped inside the nitride film,

(a) (b)

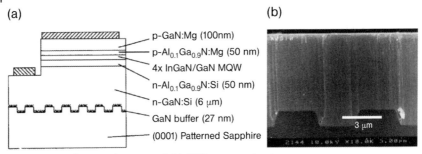

p-GaN:Mg (100nm)

p-Al$_{0.1}$Ga$_{0.9}$N:Mg (50 nm)

4x InGaN/GaN MQW

n-Al$_{0.1}$Ga$_{0.9}$N:Si (50 nm)

n-GaN:Si (6 μm)

GaN buffer (27 nm)

(0001) Patterned Sapphire

3 μm

Fig. 1.14 (a) Schematic diagram and (b) SEM cross-sectional micrograph of a 382 nm InGaN/GaN MQW LED grown on a patterned sapphire substrate. After Ref. [83].

one of the most effective ways is to texture the sapphire surface to suppress the guided modes. Figure 1.14 depicts a schematic diagram and scanning electron microscope (SEM) cross-sectional micrograph of a 382 nm InGaN/GaN MQW LED grown on a patterned sapphire substrate [83]. Parallel grooves were fabricated using standard photolithography and subsequent reactive ion etching along the $\langle 11\bar{2}0 \rangle$ direction. The flipchip-mounted LEDs exhibited an output power of 15.6 mW and an external quantum efficiency of 24% at 20 mA. The improved efficiency is a result of enhanced light extraction through the sapphire substrate as well as improved material quality yielded by lateral growth on the patterned substrate.

1.3
LEDs on SiC Substrates

SiC has also been used as a growth substrate for high-brightness InGaN-based LEDs [89, 90]. One main advantage of using SiC is that it can be easily doped to make a conductive substrate, enabling the fabrication of LEDs with a vertical geometry, similar to conventional III–V LEDs grown on conductive GaAs and GaP substrates. Nonuniform current spreading, as exists in lateral LEDs on sapphire, is therefore mitigated. The fabrication and packaging of such a LED chip is relatively straightforward, as a mesa structure is not needed. A Ni-based contact is formed on the backside of the substrate as the n-type electrode, and only one wire bond is required for top-emitting LEDs. SiC also has the advantage of being highly thermally conductive. Its thermal conductivity is ~10 times higher than that of sapphire, greatly enhancing heat removal from LED chips. Additionally, unlike sapphire, SiC has natural orthogonal cleavage planes, which facilitate the chip separation process. The cleavage planes are particularly important for producing high-quality facets to form resonant cavities in laser diodes (LDs).

Using SiC substrates for III–nitride LEDs, however, is less common mainly due to their high cost. Even though SiC has a smaller lattice mismatch (~3.5%) with GaN than sapphire, it is sufficiently large to result in a large number of dislocations (10^8–10^{10} cm^{-2}) in epilayers [91–93]. As compared to sapphire wafers, epi-ready SiC wafers typically have a rougher surface finish and a larger number of surface and bulk defects, which are additional sources of microstructural defects in epilayers. Another drawback is that doped SiC is typically opaque at the wavelengths of visible light. It is strongly absorbing at UV wavelengths shorter than its band-edge emission (~410 nm for 6H-SiC) and is not suitable for developing high-efficiency deep-UV emitters.

GaN epilayers are usually grown on the (0001) plane of Si-face 6H- or 4H-SiC substrates using the MOCVD technique. An AlN or AlGaN buffer layer is required to help accommodate the lattice mismatch and relieve stress. AlN can nucleate on SiC at both low and high temperatures [91–93]. In contrast to low-temperature buffer layers on sapphire, AlN deposited on SiC at high temperatures (~1100 °C) has better crystalline quality and surface morphology [92], and therefore it is more effective for promoting subsequent 2D growth of high-quality films. Because SiC has a smaller thermal expansion coefficient, upon cooling down from the growth temperature GaN epilayers may be under a tensile stress, which favors the formation of cracks. It has been suggested that the strain state of GaN epilayers is determined by their growth mode, and tensile strain can be avoided by growing GaN on a thin coherently strained AlN buffer layer [94].

Cree's first-generation InGaN-based blue LEDs employed a highly-resistive AlN buffer layer [89]. To bypass this resistive layer, a shorting ring was employed that allowed vertical flow from the LED structure to the SiC substrate. The AlN buffer was later replaced by a Si-doped low-Al content AlGaN layer [89], which is conductive but may still introduce a significant series resistance. State-of-the-art blue and green LEDs grown on SiC have a layer structure similar to those grown on sapphire, and are comparable in microstructural quality, electrical properties, and optical characteristics.

SiC has a refractive index (~2.7 at blue wavelengths) slightly higher than GaN, but considerably higher than epoxy. No total internal reflection occurs at the GaN/SiC interface. Thus, in LEDs with a rectangular geometry, most light emitted by the active region is trapped inside the SiC substrate and has a good chance to be absorbed by the substrate and bottom contact. Chip shaping and texturing are, therefore, critical for light extraction. A truncated inverted pyramid (TIP) LED is an example of chip shaping of an individual LED [95] that is achieved by dicing a chip with a beveled blade to yield a sidewall angle in the range of 30°–50° with respect to the vertical. Such a shape enhances light extraction by redirecting the internally reflected light and reducing the average light path within the chip, as illustrated in Fig. 1.15(a). Figure 1.15(b) shows the calculated light extraction efficiency of a blue LED on SiC as a function of the sidewall angle. An absorption coefficient of 5 cm^{-1} was used for the SiC substrate. With an angle ~35°, light extraction can be increased by 110% compared to a rectangular chip. Cree's XB LEDs employ this chip shaping technique [see Fig. 1.15(c)] in conjunction with flipchip packaging, providing a record performance with an external quantum efficiency of ~45% at 450–455 nm, corresponding to a radiant power of 25 mW at 20 mA [96].

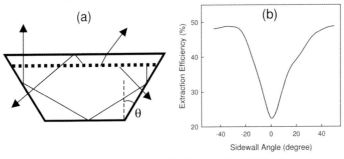 

Fig. 1.15 (a) Schematic diagram of a TIP LED illustrating the enhanced light extraction efficiency. (b) Calculated light extraction efficiency of a blue LED on SiC as a function of the sidewall angle. (c) A shaped Cree LED chip.

1.4
LEDs on Si Substrates

The allure of low-cost light sources has fueled a variety of efforts to grow III–nitride LEDs on Si substrates [97–103]. Si is thermally stable under typical nitride growth conditions. Si wafers, with high crystalline quality and a smooth surface finish, are available in large sizes at low prices. Using Si substrates allows easy die separation, simple wet-etch substrate removal, and the potential to integrate LEDs with Si control electronics.

III–nitride blue and UV LEDs have been grown on Si(111) using molecular-beam epitaxy (MBE) [97, 98] and MOCVD [99–103]. The enormous 17% lattice mismatch of GaN on Si mandates the use of a buffer layer such as AlN or AlGaN [97–108]. Besides reducing crystalline defects, the buffer layer also enhances 2D growth of GaN and prevents the formation of an amorphous SiN_x interlayer [105]. The greater thermal expansion coefficient of GaN as compared to that of Si gives rise to a tensile stress in epilayers. Cracking may occur upon cooling down from the growth temperature [99]. The cracking problem may be reduced by optimizing the structure and growth of the buffer layer. It was recently found that MOCVD-grown GaN epilayers on a thick AlN–GaN graded buffer were under a compressive stress and free of cracks [106]. Both a step-graded AlGaN buffer and an AlN/GaN superlattice have also proven to be effective for reducing the crack density [107, 108]. Most recently, by employing a high-quality AlN buffer layer, Li et al. [103] successfully scaled blue LEDs on Si wafers up to 6 inches.

LEDs grown on Si are fabricated in a similar fashion to LEDs on SiC by forming the n-type contact on the backside of the substrate and have a vertical structure, which greatly enhances current spreading uniformity. However, most reported LEDs on Si have a high series resistance, due to the large band offset between the AlN buffer and Si substrate, as well as poor doping in the p-type cladding layer [99, 102]. Despite the use of an AlN buffer layer, the crystalline quality of epilayers on Si cannot be compared to those obtained on sapphire. State-of-the-art LEDs on Si are at least 10 times less efficient than their counterparts on sapphire or SiC substrates. The highest reported performance is still sub-milliwatt output power at 20 mA and less than 1% external quantum efficiency [100]. The LED performance is fundamentally limited by two key factors: poor light extraction due to the inherent visible and UV light absorption of silicon, and high defect density due to the aforementioned high thermal and lattice mismatch between Si and nitrides. Nevertheless, with a significant cost reduction possible from 6 inch Si wafers, these LEDs hold promise for special applications where high brightness is not required.

Light absorption by the substrate results in a significant light loss in LEDs grown on Si wafers. One way to reduce this parasitic loss is to insert an AlGaN/GaN distributed Bragg reflector (DBR) between the LED structure and the Si substrate [101]. The DBR, however, will increase the operation voltage and substantially increase the possibility of film cracking. Another way is to remove the Si substrate and transfer the LED film to a reflective substrate. A similar approach has been employed to improve heat dissipation and light extraction of blue LEDs on sapphire [109–110]. In this case, the separation was realized using a laser lift-off technique due to the chemical inertness of sapphire. Si can be readily removed by selective chemical wet etching. The substrate removal can not only greatly improve the light extraction efficiency but also eliminate the conduction barrier at the buffer/Si interface. Zhang et al. [102] reported the separation of an InGaN MQW green LED from Si and transfer to a copper carrier. The LED wafer with an Al/Au metal reflector deposited on the p-contact was first bonded onto an Au plating copper carrier using indium. The Si substrate was thinned down to 60 μm and selectively etched in a $HF:HNO_3:CH_3OOH$ (1:1:1) solution. The AlN/AlGaN buffer layer was then plasma etched, and the n-type electrode, which consists of a Ti/Au (5 nm/5 nm) semitransparent contact and an Al/Au bondpad, was formed on the exposed n-GaN layer. Figure 1.16 shows the *I–V* and *L–I* characteristics of the LED before and after this process. The series resistance was reduced from 42 Ω to 27 Ω, and the optical power was increased by 49%. Further improvement in light extraction can be expected by using a more transparent n-type contact and an appropriate encapsulation.

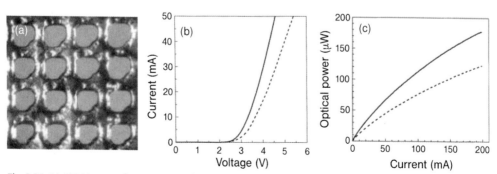

Fig. 1.16 (a) SEM image of a green LED after partial removal of the Si substrate. (b) The *I–V* characteristics and (c) the *L–I* characteristics of the LED before (dashed line) and after (solid line) substrate removal. After Ref. [102].

1.5
LEDs on Free-Standing GaN Substrates

For LEDs grown on foreign substrates, the mismatches in lattice constant and thermal expansion coefficient between the nitrides and substrates manifest themselves as a high density of threading dislocations and residual biaxial stress in the epilayers. High-quality bulk GaN substrates, if available, would significantly mitigate these problems. The growth procedure for GaN-based LEDs can be greatly simplified as the homoepitaxy process does not require additional steps, such as surface nitridation and low-temperature buffer layer, which are mandatory in heteroepitaxial growth. The defects and stress in the epilayers would be greatly reduced, leading to improved performance, yields, and scalability to larger substrates. In addition, simple vertically structured LEDs can be fabricated on conducting GaN substrates. The vertical geometry, in combination with good thermal conductivity of GaN (5× higher than sapphire), allows the LEDs to operate at much higher current densities and temperatures.

1.5.1
LED Homoepitaxy

The chemical passivity of nitrogen, and the high melting temperature and high decomposition pressure of GaN have made growth of large-size and high-quality GaN boules problematic. A number of bulk GaN crystal solution growth techniques are under development [111–115], which use liquid gallium or gallium alloys as a solvent and a high pressure of nitrogen above the melt to maintain GaN as a thermodynamically stable phase. One of the most intensively explored approaches is high-pressure solution growth (HPSG) developed by Porowski et al. [111, 112]. Bulk GaN crystals are grown from a Ga melt, under a nitrogen hydrostatic pressure of about 15 kbar at temperatures ranging from 1500 to 1800 K, and crystallize in the form of platelets or rods. The process is capable of growing GaN crystals with a dislocation density lower than $10^4 \, cm^{-2}$. The undoped crystals have a high n-type background doping, on the order of $5 \times 10^{19} \, cm^{-3}$, which is believed to be due to oxygen impurities and nitrogen vacancies, which also causes some crystal opacity. One major drawback of such a technique, however, is that the quality of the GaN crystals deteriorates with increasing growth rate and size, thus limiting the maximum crystal size to ~10 mm.

A more mature technology for growing bulk GaN is hydride vapor-phase epitaxy (HVPE) [116–121], which is normally carried out in a hot wall reactor at atmospheric pressure. In this approach, vapor-phase GaCl, formed by reacting HCl with liquid Ga at 800–900 °C, is transported to a substrate, such as sapphire, Si or GaAs, where it reacts with injected NH_3 at 900–1100 °C to form GaN. As compared to other epitaxial growth approaches, much higher growth rates (on the order of $100 \, \mu m \, h^{-1}$) can be attainable by HVPE [119], making it a potentially low-cost technique for mass-producing bulk GaN. Free-standing GaN substrates can be produced by separating thick HVPE grown GaN from the underlying substrate using laser beam radiation or mechanical polishing. Lapping, polishing, and etching are then performed to achieve GaN wafers with an epi-ready surface.

The dislocation density in HVPE-grown GaN films is initially quite high, on the order of 10^{10} cm^{-2} as is typical for heteroepitaxially grown thin GaN layers. As the epilayer thickness increases, the evolution and annihilation of the dislocations lead to a reduction in dislocation population. The density usually drops to a value of about 10^7 cm^{-2} after a thickness of 100–300 µm of GaN has been grown [119]. However, large strain and associated bowing or cracking, which result from the use of a foreign substrate, will limit the maximum film size and thickness. The strain and bowing remain even after removal of the original substrate, and are expected to be also present in epitaxial layers deposited on such substrates. By improving strain management, Xu et al. [120] successfully fabricated 2 inch GaN boules as thick as 10 mm with a dislocation density as low as 10^4 cm^{-2}. HVPE GaN may also contain a high density of point defects, such as N vacancies and impurities, leading to a high background doping level and a redshift of the emission spectrum.

GaN homoepitaxy using MOCVD is more straightforward than heteroepitaxial growth. The substrate is heated directly to the growth temperature in an ammonia-rich environment. 2D growth mode can be achieved on GaN(0001) without using a buffer layer. Homoepitaxially grown GaN usually shows a smooth surface with a step structure, in contrast to a swirled step structure for typical epitaxy on sapphire. It has been found that vicinal surfaces with an offcut of 1–2 degrees yielded smoother morphology than nominal *c*-plane [121]. This can be understood by the fact that the offcut surfaces provide predefined atomic steps for smooth 2D step-flow growth. Homoepitaxial GaN, in most cases, replicates the defect structure in the bulk substrates [122]. However, surface defects introduced in the polishing and etching steps were found to have a pronounced impact on the homoepitaxial quality [123] and must be removed by additional chemomechanical polishing or chemically assisted ion-beam etching. Figure 1.17 shows high-resolution X-ray diffraction (HRXRD) rocking curves of the (0002) reflection of an HVPE GaN substrate, and GaN epilayers grown atop the substrate and sapphire. The FWHM of the homoepitaxial GaN is 79 arcsec, which is comparable to that of the substrate (85 arcsec), but much smaller than for the heteroepitaxial GaN (230 arcsec). Since the FWHM of the (0002) peak reflects the degree of lattice distortion from dislocations, the smaller FWHM of the homoepitaxial GaN confirms that the threading dislocation density is substantially reduced. Our recent study showed that the incorporation of common impurities in homoepitaxial GaN, including C, H and O, was significantly reduced compared to GaN grown on sapphire [124]. This may be a result of dislocation reduction in the homoepitaxial GaN as impurities tend to congregate around microstructural defects and create localized states in the bandgap [125].

The growth of III–nitride LEDs on bulk GaN substrates is referred to here as homoepitaxy even though the alloy composition and lattice constants of the active and cladding layers may not be identical to those of the substrate. While visible LEDs with GaN cladding layers are well matched to the substrate and slightly

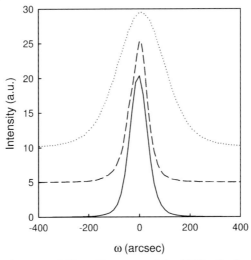

Fig. 1.17 HRXRD rocking curves of the (0002) reflection from an HVPE GaN substrate (solid line) and GaN epilayers grown on bulk GaN (dashed line) and sapphire (dotted line).

strained, UV LEDs based on AlGaN heterostructures grown on bulk GaN may suffer from a large tensile stress, thus causing cracks. An additional stress-relief layer, such as a graded AlGaN layer, is required for growing thick LED structures.

Pelzmann et al. [126] reported the growth of homojunction GaN LEDs by depositing a p-GaN layer on an n-type HPSG GaN substrate. The homoepitaxial LEDs demonstrated a doubling of the emission intensity relative to their counterparts on sapphire. Franssen et al. [127] studied the electrical and optical characteristics of HVPE-grown InGaN/GaN MQW blue LEDs on a similar GaN substrate. Thermionic emission rather than tunneling transport was found to be the main mechanism responsible for radiative recombination in the LEDs. The maximum internal and external quantum efficiencies were determined to be 38% and 1.9%, respectively.

We have investigated the growth of InGaN/GaN MQW LEDs on free-standing HVPE GaN substrates using low-pressure MOCVD. The GaN substrates, offcut by 1–2 degrees in the $\langle 11\bar{2}0 \rangle$ direction, were ~300 µm thick, unintentionally doped, and had a free carrier concentration of ~7×10^{17} cm^{-3}. Atomic force microscopy (AFM) measurement showed that threading dislocations were terminated by surface pits after the chemomechanical polish and the density was determined to be ~$(1–2) \times 10^{7}$ cm^{-2}. The substrates were heated directly to 1050 °C in a steady ammonia flow, followed by the growth of the n-GaN cladding layer. The growth recipes for subsequent InGaN QWs and AlGaN layer were similar to those known for heteroepitaxial LEDs. The active regions were 10-period InGaN/GaN MQWs with an In composition ranging from 0.1 to 0.21.

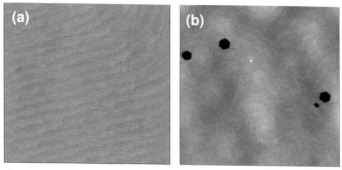

Fig. 1.18 AFM images (2 μm × 2 μm) of InGaN/GaN MQW LEDs grown on (a) GaN and (b) sapphire.

Figure 1.18(a) shows a 2 μm × 2 μm AFM image of the homoepitaxial LED. The surface is defect-free, and a step structure with terraces of ~100 nm is clearly seen. In contrast, a similar structure grown in the same epitaxy run on sapphire with a previously grown buffer exhibits a rougher morphology and ~1 × 10^8 cm^{-2} V-defects ranging from 50 to 150 nm [see Fig. 1.18(b)]. The rms roughness of the LEDs on GaN and sapphire are 0.23 nm and 0.65 nm, respectively. The smooth surface of the homoepitaxial LED suggests abrupt heterostructural interfaces and uniform QWs resulted from 2D step-flow epitaxy. The V-defects, which are open hexagonal inverted pyramids defined by six $\{10\bar{1}1\}$ planes, are connected to threading dislocations mostly with screw and mixed character [128]. Microstructural defects and residual stress, which promote the formation of V-defects, are substantially reduced in the homoepitaxial LED. We have found that the growth temperature for the LED on GaN was slightly higher at the same epitaxy conditions, presumably due to the higher thermal conductivity of the substrate. This may partly account for the suppression of V-defects by enhancing the rates of Ga diffusion and incorporation on the off-axis facets [26, 129]. Cross-sectional TEM measurements showed that the density of threading dislocations reaching the active region in the LED on sapphire was ~2 × 10^9 cm^{-2}, whereas it was much lower in the homoepitaxial LED and could not be precisely determined.

1.5.2
Electrical Characteristics

In contrast to lateral LEDs fabricated on sapphire, the LEDs on GaN, with their n-contact formed on the backside of the substrate, have a vertical structure. Figure 1.19 compares the forward *I–V* characteristics of two representative LEDs at various temperatures. As expected, the LED on sapphire shows a characteristic tunneling behavior in the forward direction. Two main exponential segments with different slopes are distinguished at low and moderate forward biases, and can be described by Eq. (1.4). The energy parameter *E*, which is temperature-insensitive,

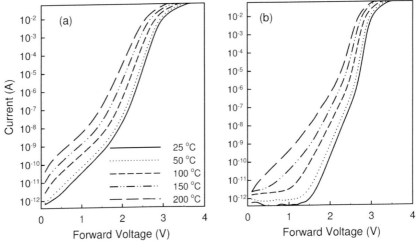

Fig. 1.19 Temperature-dependent forward *I–V* characteristics of InGaN/GaN MQW LEDs grown on (a) sapphire and (b) GaN.

has a value of 190 meV at 0–2.0 V and 70 meV at 2.0–2.8 V. No realistic ideality factors can be extracted. The forward *I–V* characteristics of the LED on GaN, as seen in Fig. 1.19(b), also divide into two distinct linear sections with different slopes, which, however, appear to be a function of temperature. Tunneling may still dominate at low injection levels, but the slope change suggests the involvement of a thermally activated current. As the forward current increases, diffusion and recombination currents start to dominate over the tunneling component. The forward current at bias >2.6 V can be described by Eq. (1.3) with an ideality factor $n = 1.5$ until series resistance in the diode dominates. The presence of the diffusion–recombination currents reflects the high material quality of the homoepitaxial LED, where defect-assisted tunneling current is greatly suppressed.

Carrier tunneling is also seen in the LED on sapphire under reverse bias, as suggested by the strong field dependence but low temperature sensitivity of the reverse leakage current shown in Fig. 1.20(a). In contrast, the LED on GaN shows a dramatic reduction in reverse current by more than six orders of magnitude (Fig. 1.20(b)). The remaining leakage current is a function of both applied bias and temperature, suggesting the coexistence of tunneling and thermal generation currents [45]. In particular, as temperature increases from 100 to 150 °C, there is a sudden jump in the low-bias current, probably arising from thermal ionization of carriers from deep traps and a trap-assisted tunneling process. An activation energy of ~0.6 eV is extracted from the $\log I$ versus $1/T$ plot at −12.5 V where the temperature dependence of the current is roughly an exponential function.

To investigate the microstructural origin of the observed difference in the electrical characteristics of the LEDs, the unmetalized portion of the sample surface was characterized using AFM and conductive AFM (C-AFM). Data were collected

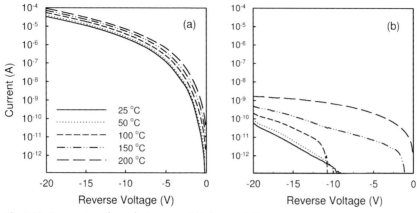

Fig. 1.20 Temperature-dependent reverse *I–V* characteristics of InGaN/GaN MQW LEDs grown on (a) sapphire and (b) GaN.

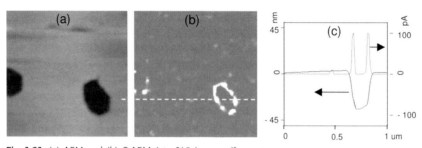

Fig. 1.21 (a) AFM and (b) C-AFM (at −3 V) images (1 μm × 1 μm) of an InGaN/GaN MQW LED on sapphire. (c) Dual section profile taken along the dashed line indicated in (b). The gray scales are (a) 10 nm and (b) 100 pA. The positive current in (c) corresponds to a reverse-bias current in the LED.

under ambient conditions using a conducting diamond-coated tip on a Veeco Instruments Dimension 3000 microscope. A positive or negative bias was applied to the n-type ohmic contact, while holding the tip at ground potential in contact to the topmost p-GaN layer to map the surface current flowing through the junction [128]. The topography of the p-GaN was recorded simultaneously.

The surface of the homoepitaxial LED is free of defects and no current was detected in this sample within the detection limit of our C-AFM system. Figures 1.21(a) and (b) show the topography and current maps of the heteroepitaxial LED over a 1 × 1 μm² area. The C-AFM current image recorded at −3 V reverse bias (applying a +3 V voltage to the n-ohmic contact) reveals nanoscopic conductivity of the LED structure on sapphire, which correlates well with the topography image. It is striking that the leakage current is highly localized at the edge of the V-defects.

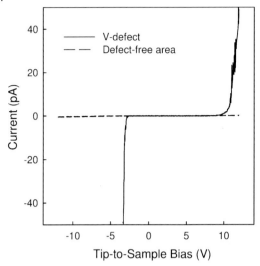

Fig. 1.22 Nanoscale local *I–V* characteristics of an InGaN/ GaN MQW LED on sapphire recorded at a V-defect and in a defect-free region.

A dual section profile [Fig. 1.21(c)] shows that the current value is ~100 pA, which corresponds to a current density of ~10 A cm^{-2}. The current scan was repeatable at the same defect, indicating that there was no significant charge accumulation. Increased current conduction was also seen along the edge of the V-defects under forward bias, though the current was much lower. These results are consistent with previous findings that mixed and screw dislocations in GaN are electrically active, creating discrete current paths [130, 131]. It is therefore clear that the V-defects and associated dislocations in the LED on sapphire act as leakage paths connecting the p-GaN and n-GaN layers, and are mainly responsible for the high reverse current observed in this device.

Figure 1.22 presents the nanoscale regional *I–V* characteristics of the LED on sapphire. The data were recorded by placing the diamond tip at a V-defect or at a fixed defect-free region while ramping the bias applied to the n-ohmic contact from +12 V to −12 V. Similar to the case of the homoepitaxial LED, the current in the defect-free region is negligible, whereas the defective area exhibits abrupt turn-on characteristics in both directions. Interestingly, the current rise takes place at a much lower voltage under reverse bias than forward bias. This is because in the latter case the tip behaves as a nanoscale Schottky contact to p-GaN under a reverse bias [132], and therefore the actual voltage dropped on the LED junction may be much smaller than the voltage applied. Given the asymmetric *I–V* characteristics of the reverse-biased nanoscale Schottky contact, the *I–V* curve at the V-defect shown in Fig. 1.22 appears to contain a symmetric current component. Indeed, a similar symmetric leakage current can also be identified near zero bias in Figs. 1.19(a) and 1.20(a). These results support the assumption that the dislocations associated with the V-defects behave as small shunt resistors connected across the p–n junction.

1.5.3
Optical Characteristics

Figure 1.23 shows a series of EL spectra of near-UV LEDs on GaN and sapphire at different injection levels. The presence of Fabry–Pérot interference fringes is a characteristic feature for LEDs grown on sapphire. As expected, they are absent in the homoepitaxial LED. The peak wavelength of the LED on GaN is 405.5 nm, whereas it is ~5 nm longer on sapphire. This discrepancy may be attributed to the slightly different growth temperatures of the LEDs arising from the difference in substrate thermal conductivity and thermal coupling of the substrates to the sus-ceptors. Neither LED shows a significant blueshift in emission energy with increas-ing current, as is commonly observed in blue and green LEDs, suggesting weaker localization effects due to a smaller In content. At high currents (>200 mA), the peak wavelength of the LED on sapphire exhibits a redshift as a result of severe joule heating. Another interesting feature shown in Fig. 1.23 is that the spectra of the homoepitaxial LED are narrower (FWHM ~16 nm). This is likely due to smaller composition or thickness fluctuation of the homoepitaxial QWs.

Figure 1.24(a) presents the *L–I* characteristics (on a log–log scale) of the near-UV LEDs measured in continuous-wave (CW) injection mode. The light output of the homoepitaxial LED increases steadily with increasing current, and the dependence

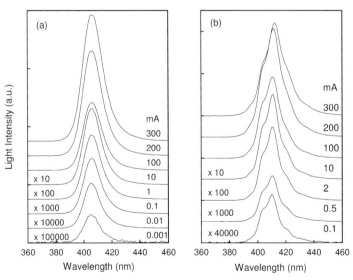

Fig. 1.23 EL spectra of near-UV LEDs on (a) GaN and (b) sapphire at different injection currents.

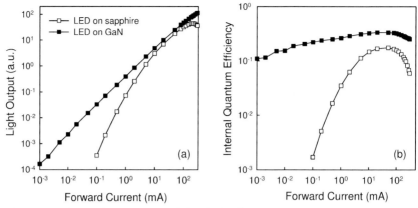

Fig. 1.24 (a) The *L–I* characteristics and (b) calculated internal quantum efficiencies of near-UV LEDs on GaN and sapphire.

is nearly linear over the entire current range. In contrast, the light output of the LED on sapphire shows a superlinear increase at low currents, and a rollover at ~180 mA. At 0.2 mA, 20 mA, and 200 mA, the respective output powers of the LED on GaN are 35×, 56% and 104% higher than the LED on sapphire. The remarkable increase in light emission at low injection levels can be ascribed to a reduced defect density, thus a lower nonradiative recombination rate, as well as enhanced carrier confinement due to higher structural quality of the confining layers. At high currents, the defect states are saturated. The superior performance of the LED on GaN in this case is due largely to the improved current spreading and heat dissipation.

To deduce the internal quantum efficiencies of the LEDs from the measured optical powers, a ray-tracing simulation was performed to determine the light extraction efficiencies. The absorption coefficient of the GaN substrate at 405 nm was found to be ~12 cm^{-1} as determined by transmission measurements, whereas sapphire is essentially transparent. The below-bandgap absorption in the HVPE GaN is caused by a large amount of defects and impurities incorporated during the crystal growth. Like LEDs on SiC, LEDs on GaN have a very small escape cone in all directions due to the high refractive index of GaN ($n \sim 2.5$). As a result, a considerable amount of light generated in the MQWs is trapped and absorbed by the substrate. The calculated chip-to-air light extraction efficiencies of the LEDs on GaN and sapphire at 405 nm are 8.2% and 11.5%, respectively. Taking these into account, the respective increases of the internal quantum efficiency at 0.2 mA, 20 mA and 200 mA are 46×, 103% and 165%. Figure 1.24(b) illustrates the calculated internal quantum efficiency (η_i) of the LEDs as a function of injection current. At low currents, η_i increases rapidly with increasing current as the radiative recombination rate increases. The efficiency of the UV LEDs reaches its maximum value of 33.7% on GaN and 16.8% on sapphire in the current range of

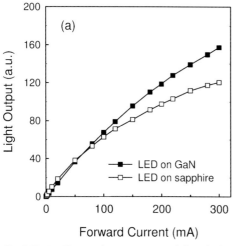
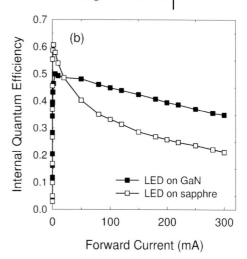

Fig. 1.25 (a) The *L–I* characteristics and (b) calculated internal quantum efficiencies of blue LEDs on GaN and sapphire.

20–50 mA. The peak efficiency of the homoepitaxial LED is expected to be further improved by optimizing its structure design and growth condition. As the current increases further, η_i decreases due to the occurrence of current overflow as the bands are filled up. The efficiency of the homoepitaxial LED decreases at a lower rate at high currents, indicating its suitability for high-power operation.

It is expected, however, that the improvement in material quality, as yielded by homoepitaxy, will have a smaller impact on the optical performance of blue and green LEDs due to stronger localization effects. Figure 1.25(a) compares the *L–I* characteristics of two blue LEDs grown on different substrates. The homoepitaxial LED significantly outperforms its counterpart on sapphire only at high currents (>50 mA), due largely to more efficient heat removal through the GaN substrate. At 20 mA, the light output of the LED on GaN is actually slightly lower than the LED on sapphire. The light extraction efficiencies of the LEDs on GaN and sapphire at 470 nm were calculated to be 9.5% and 12.1%, respectively. The internal quantum efficiencies of the two LEDs at 20 mA are therefore comparable, ~49%, as seen in Fig. 1.25(b). The efficiency of the blue LEDs peaks at a much smaller current of 5 mA as compared to the near-UV LEDs. It then decreases sharply with increasing current for the blue LED on sapphire. This cannot simply be interpreted as a self-heating effect, and may be associated with inefficient carrier capture by localized states due to poor microstructural properties of the heteroepitaxial LED.

As discussed earlier, the performance of AlGaN-based UV LEDs grown on sapphire is limited by poor material quality. Substantial reduction in dislocation density can be achieved by using lattice-matched GaN substrates. It is anticipated that efficient UV LEDs with peak wavelength shorter than 365 nm can be fabricated on bulk GaN despite strong light absorption by the substrate. Nishida et al. [133] demonstrated 352 nm $Al_{0.04}Ga_{0.96}N$ SQW LEDs on a HVPE GaN substrate with 0.55 mW output power and 1% external quantum efficiency at 20 mA. A short-period $Al_{0.16}Ga_{0.84}N/Al_{0.2}Ga_{0.8}N$ superlattice was employed as the transparent p-type cladding layer in the top-emitting LEDs. However, growing thick AlGaN cladding layers on GaN becomes increasingly difficult with increasing Al content due to the lattice and thermal mismatch between AlN and GaN. By using a stress-relief template consisting of 25-period n-type $Al_{0.25}Ga_{0.75}N/Al_{0.2}Ga_{0.8}N$ superlattices with a period of 4 nm, Yasan et al. [134] were able to grow a 340 nm LED structure on a free-standing GaN. The device showed a reduced series resistance and improved output power that is one order of magnitude higher than that of similar LEDs grown on sapphire.

To gain insight into the impact of microstructural defects on the carrier dynamics in AlGaN materials, Garrett et al. [135] performed time-resolved photoluminescence (TRPL) characterization of eight-period $Al_{0.1}Ga_{0.9}N(3\,nm)/Al_{0.3}Ga_{0.7}N(7\,nm)$ MQW structures grown on GaN and sapphire substrates. The MQW structure on GaN was fully strained and under a tensile stress. The PL emission centered at 340 nm was four times more intense than from the MQWs on sapphire. Figure 1.26 shows a comparison of TRPL decays for the AlGaN MQWs for the same pump intensity and pump pulse centered at 275 nm. The transient PL for the MQWs grown on sapphire is characterized by a room-temperature lifetime of ~120 ps and a slow decay (~150 ps) that dominates at longer times. The MQWs on GaN exhibit a much longer lifetime, ~500 ps, and a longer slow decay time, ~488 ps. The longer

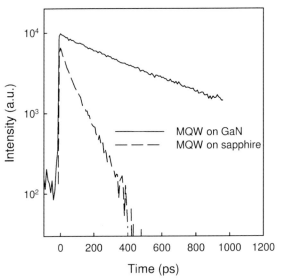

Fig. 1.26 TRPL decay curves at the same pump intensity at 275 nm for $Al_{0.1}GaN/Al_{0.3}GaN$ MQWs grown on GaN and sapphire. After Ref. [135].

PL lifetime of the AlGaN MQWs on GaN reflects better material quality and an increase in nonradiative lifetime associated with the reduction in dislocation density by about two orders of magnitude.

1.5.4
High-Current Operation

In order to compete with fluorescent and other conventional lighting sources, it is essential that the cost of LEDs is further reduced and their efficiency is improved. One way to meet the cost and performance targets is to drive LEDs at much higher current densities without compromising emission efficiency and operating life-time [136]. III–nitride LEDs grown and fabricated on sapphire are not suitable for high-power operation for several reasons. First, LEDs grown on sapphire contain a high density of threading dislocations, which may accelerate device degradation particularly at high currents. Second, due to the insulating substrate, LEDs on sapphire are normally fabricated in a lateral device configuration. Mismatch between the n-type and p-type current spreading layers may cause severe current crowding and localized self-heating [32–34]. Finally, sapphire has a rather poor thermal conductivity, limiting heat dissipation and, therefore, the maximum oper-ational temperature and power. Homoepitaxial LEDs would overcome all these drawbacks. High material quality, along with the good thermal and electrical con-ductivity of the substrate, makes homoepitaxial LEDs more suitable for high-current and high-temperature operations.

To examine the impact of device geometry on high-current performance, near-UV LEDs with both lateral and vertical structures were fabricated on HVPE GaN. The chip size was 300 μm × 300 μm. The top contact pattern and schematic cross-section of the LEDs are shown in Fig. 1.27. Some p-type finger projections were

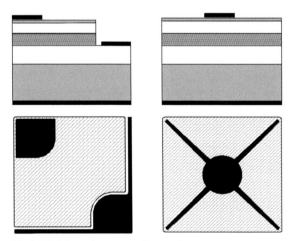

Fig. 1.27 Schematic cross-sections and top contact patterns of lateral and vertical LEDs on sapphire or GaN substrates.

added to the bondpad to alleviate current crowding by reducing the current spreading distance. The vertical LED has an emitting area ~20% larger than the lateral LED because the mesa structure is not needed. The lateral device has a conventional asymmetric structure similar to the LED fabricated on sapphire. Using Eq. (1.1), we calculated the current spreading length L_s in the LED on sapphire to be 250 μm for the measured values of $\rho_p = 2.7\,\Omega\,cm$, $t_p = 0.2\,\mu m$, $\rho_n/t_n = 26\,\Omega$, $\rho_t/t_t = 18\,\Omega$, and $r_c = 5 \times 10^{-3}\,\Omega\,cm^2$. Current crowding takes place at mesa edges and the current density drops to ~1/e under the p-type bondpad. On the contrary, current tends to crowd toward the p-type bondpad in the lateral LED on GaN due to the thick conductive substrate. Uniform current spreading may be achieved by optimizing the transparent contact to meet $\rho_t/t_t = \rho_n/t_n$. However, due to the different temperature dependences of ρ_t and ρ_n, current crowding may occur as the junction temperature increases at high operating currents.

The vertical LED has a symmetric structure and contact geometry similar to conventional AlInGaP LEDs grown on conductive GaAs. Current spreads uniformly under the p-type bondpad metal. The current spreading length is independent of the conductivity of the n-GaN layer and varies approximately as $(\rho_t/t_t)^{1/2}$. By adding the cross-shaped fingers, lateral current paths in the transparent contact are considerably reduced, leading to a nearly uniform current distribution and light emission.

Figure 1.28 compares the forward *I–V* characteristics of the lateral and vertical LEDs. The series resistances of the vertical LED on GaN and the lateral LEDs on GaN and sapphire are $7\,\Omega$, $12.2\,\Omega$, and $14.2\,\Omega$, respectively. The high series resistance of the lateral devices is due largely to an extra voltage drop within the current

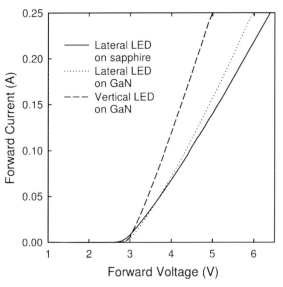

Fig. 1.28 Forward *I–V* characteristics of lateral and vertical LEDs on sapphire or GaN.

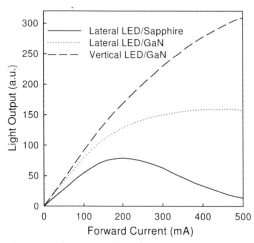

Fig. 1.29 Light output of lateral and vertical LEDs on sapphire or GaN as a function of injection current measured in CW mode.

spreading layers. The forward voltage of the vertical LED at 200 mA is 4.5 V, much lower than 5.8 V for the lateral LED on sapphire. With a much lower series resistance, and therefore a reduced thermal load, the vertical LED has a higher allowable operating temperature and power conversion efficiency.

The output power of the LEDs as a function of injection current measured in CW mode is shown in Fig. 1.29. The homoepitaxial LEDs greatly outperform the LED on sapphire at high injection levels. The output power of the lateral LEDs on sapphire and GaN peaks at ~200 mA and 400 mA, respectively. The power rollover can be attributed to severe current crowding and self-heating effects, which are even more pronounced in the LED on sapphire due to the poor thermal conductivity of sapphire. As the LED junction temperature increases, carrier confinement in the MQWs becomes less efficient, leading to a decreased radiative efficiency. In sharp contrast, the output power of the vertical LED on GaN increases steadily with increasing current and shows no saturation up to 500 mA. At this current, the vertical LED chip has a power conversion efficiency two times higher than the lateral LED on GaN and 28 times higher than the LED on sapphire. These results illustrate the critical need to develop efficient thermal management schemes for high-power LED packages.

LED reliability at high currents was evaluated by stressing the LEDs at 400 mA, which is 20 times higher than standard rated current for most commercially available blue LEDs of a similar size. The average current density in the lateral LEDs was 620 A cm^{-2}, and localized current densities could be much higher due to non-uniform current spreading. Figure 1.30 shows the variation of light output as a function of stress time. The optical power of the vertical LED is essentially unchanged (<1% decrease), suggesting excellent reliability. The lateral LED on

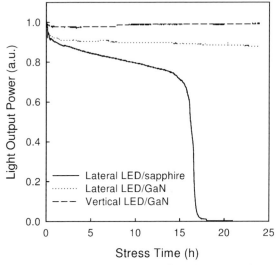

Fig. 1.30 Normalized optical power of lateral and vertical LEDs on sapphire or GaN as a function of stress time. The LEDs were stressed at 400 mA at room temperature.

GaN exhibits a gradual degradation and a 12% decrease after 24 h stress. The light output L is roughly an exponential function of t,

$$L = L_0 \exp(-\beta t) \tag{1.9}$$

and the value of the degradation rate β is determined to be $1.9 \times 10^{-3} \, h^{-1}$. The lateral LED on sapphire also shows a gradual decay during the first 15 h, though with a much higher rate of $1.4 \times 10^{-2} \, h^{-1}$. Further stress leads to a nearly catastrophic failure. The drastic drop in light output is accompanied by the destruction of the p-type contact, suggesting that the device failure may be caused by contact degradation at elevated temperatures.

To glean further insight into the stress failure mechanism, symmetrical lateral LEDs with an n-electrode ring surrounding the mesa were also fabricated on sapphire and stressed under the same conditions. Current spreading was greatly improved in these LEDs and optical degradation was less than 5% after the stress. The slow degradation rate confirms that pre-existing dislocations ($\sim 10^9 \, cm^{-2}$) in the LEDs grown on sapphire do not drive the degradation of optical power at current density up to $\sim 700 \, A \, cm^{-2}$. Threading dislocations in GaN and its alloys have been predicted to have a much lower mobility than those in conventional III–V semiconductors, due partly to a small shear stress for dislocation motion [137].

It is worth mentioning that III–nitride laser diodes (LDs) typically operate at current densities in the kA cm^{-2} range, and therefore are more sensitive to the presence of the dislocations [138–141]. Indeed, the success of GaN-based LDs is based on the development of the lateral epitaxial overgrowth (LEO) technique [139, 140]. The dislocation density in GaN grown by LEO on sapphire can be reduced to 10^5–10^7 cm^{-2}. The lifetime of LDs grown on LEO GaN was found to be one order of magnitude longer than that of LDs grown on sapphire substrates [139]. The defect density may be further reduced by using LEO in conjunction with the HVPE technique [141]. Homoepitaxy of LD structures on such a low-defect GaN substrate should have a significant positive impact on the device lifetime, particularly at high power consumption.

1.6
LEDs on Other Novel Substrates

Other novel substrates pursued for the epitaxy of III–nitride LEDs can be categorized into two groups: those providing better lattice-constant matching for the nitrides and those producing nonpolar LED structures.

Like bulk GaN, AlN is an ideal substrate for III–nitride epitaxy. It has excellent mechanical strength and thermal stability, and closely matches the crystal structure and lattice parameter of high-Al AlGaN. These properties, along with its UV transparency and high thermal conductivity (3.2 W cm^{-1} K^{-1}), make it particularly attractive for the growth of deep-UV LEDs. LEDs on AlN substrates must be fabricated in a lateral configuration due to its insulating nature. Native AlN crystals, however, do not exist. To date, sublimation–recondensation has been the most successful technique to produce bulk AlN [142–144]. Recent progress has resulted in production of high-quality AlN crystal exceeding 1 inch in diameter and having a dislocation density less than 10^4 cm^{-2} [143, 144].

Epi-ready AlN substrates with surface roughness ~1 nm can be prepared using chemomechanical polishing [145]. Epitaxial growth on AlN substrates has been found to be limited by the affinity for Al_2O_3 formation at the AlN surface [146]. The dislocation density in AlN and AlGaN epilayers was reported to be ~1 × 10^8 cm^{-2} [145, 146], much higher than those in the AlN substrates. In situ reflection high-energy electron diffraction revealed crystalline Al_2O_3 islands on the Al face of AlN [146]. The residual oxide islands, which may form by exposure to air, lead to three-dimensional growth and an increased dislocation density in the epilayers. A relatively clean surface can be obtained by exposing AlN to an NH_3 flux to nitridate the surface [146], but a complete removal of the native oxide is needed to obtain high-quality epilayers. Fully strained $Al_{0.5}Ga_{0.5}N$/AlN MQWs have been grown on AlN using MOCVD, which demonstrated photoluminescence at 260 nm with intensity 28 times higher than a similar structure grown on SiC [147]. Nishida et al. [148] reported an AlGaN MQW LED on AlN with a peak wavelength at 345 nm. The output power linearly increased up to 300 mA for a 300 μm × 300 μm flipchip LED, and the maximum optical power was ~1.2 mW. In comparison to similar UV LEDs grown on sapphire, the LED on AlN exhibited a notably higher saturation current, which is attributed to the much higher thermal conductivity of AlN.

Other crystals that may be considered as the substrates of III–nitride LEDs include ZnO, LiGaO$_2$, and ZrB$_2$. These crystals have a crystalline structure similar to III–nitrides, and their lattice constants are closely matched. These substrates can be selectively etched, facilitating substrate removal for better heat dissipation or light extraction. One major disadvantage of using these substrates is their poor thermal stability under nitride growth conditions [149–152]. For instance, at typical GaN epitaxy temperatures, diffusion and reaction between LiGaO$_2$ and GaN may produce a noncrystalline interfacial layer, degrading the quality of epilayers [150]. InGaN/GaN MQW LEDs have been successfully grown on a ZrB$_2$ substrate with a low-temperature AlN buffer layer [152]. The buffer layer was deposited at 600 °C, which is lower than the formation temperature of an undesirable ZrBN layer.

All the substrates discussed to this point produce (0001) *c*-plane wurtzite III–nitride materials, as this orientation is favorable for smooth epitaxial growth. In *c*-plane LED structures, there is a strong piezoelectric and spontaneous polarization along the *c*-axis. The built-in electric field separates electron–hole pairs, therefore reducing the radiative recombination efficiency and causing a redshift of the emission peak. A direct strategy to eliminate the polarization effects is to grow LED structures with a nonpolar orientation, such as (11$\bar{2}$0) *a*-plane and (10$\bar{1}$0) *m*-plane or with a semipolar orientation such as (11$\bar{2}$n) and (10$\bar{1}$n) planes. The use of nonpolar-plane epitaxy allows LED growth in the direction perpendicular to the axis of polarization, whereas semipolar planes allow growth in an off-axis direction from the *c*-plane polarization vector, eliminating or reducing the polarization field in the active region.

Nonpolar LEDs have been grown on *a*-plane HVPE GaN [153], *r*-plane sapphire [152, 154, 155], as well as *m*-plane GaN [156] and 4H-SiC substrates [152, 157]. In most cases, LED performance has been limited by rough morphology, high density of dislocations and stacking faults, which arise from asymmetric off-*c*-axis growth [158]. Recently, the performance of *a*-plane LEDs was significantly improved by using a lateral epitaxially overgrown *a*-plane GaN template [153, 159]. The LEO process yielded *a*-plane GaN with an improved surface morphology and markedly reduced density of dislocations and stacking faults. In$_{0.17}$Ga$_{0.83}$/GaN MQWs LEDs grown on the template had an output power of 0.24 µW at 20 mA and current-independent emission peak at 413 nm [153]. The latter offers indirect experimental evidence that the polarization field is negligible in these LEDs. Blue LEDs grown on *m*-plane GaN showed a similar optical efficiency, but a significant current-induced blueshift in peak emission, which was attributed to band filling of localized states [156].

Polarized light emission from *a*-plane and *m*-plane LEDs has been predicted by valence band calculations of wurtzite GaN assuming a strong *c*-axis polarization [160–162]. In MQW LEDs, the interaction of the multiple valence bands with the strong *c*-axis polarization dipole supports radiative recombination at specific optical polarizations depending on the orientation of the light-emitting QWs with respect to the *c*-axis. Polarized light emission along the ⟨1$\bar{1}$00⟩ direction has been observed from an *m*-plane blue LED [157, 162]. The emitted light was partially polarized in the ⟨11$\bar{2}$0⟩ direction with a polarization ratio of 0.17 [162]. Preferential optical polarization can be useful for liquid-crystal display applications and other applications where polarized light is essential for operation.

1.7
Conclusions

Since the development of high-quality GaN and AlN buffer layers, sapphire and SiC have become the dominant substrates for III–nitride LEDs. High-brightness InGaN-based green, blue and near-UV LEDs have been demonstrated and commercialized despite the presence of a high dislocation density in the heteroepitaxial structures. Localization effects induced by In compositional fluctuation are believed to be responsible for the enhanced radiative process. The well-established growth and processing technologies for LEDs on sapphire and SiC justify the enormous ongoing efforts to develop high-efficiency and low-cost LEDs on these substrates for next-generation lighting. Making solid-state lighting a reality will require further improvements in LED efficiency and reliability, and reductions in cost. Optimizing reactor design and growth recipes, and developing novel light extraction approaches are expected to provide opportunities for breakthrough.

The low cost and high quality of Si, and the possibility of integration with Si electronics, make it an attractive substrate for III–nitride LED growth. Blue LEDs with reasonable performance have been made on up to 6 inch Si substrates despite a large mismatch between Si and GaN. It is likely that the use of Si substrates will be limited to low-cost and low-brightness light-emitting devices, unless the epitaxy quality is considerably improved and an economical film-transfer process is developed.

Sapphire will likely be the substrate of choice for deep-UV LEDs due in part to its transparency. Milliwatt UV LEDs with emission wavelengths as short as 250 nm have emerged in the past several years. Further improvement in LED performance is expected in the near future as the heteroepitaxy of AlGaN materials on sapphire is refined. AlN is a much closer lattice and thermal match for High-Al content AlGaN heterostructures. If bulk AlN or AlGaN becomes commercially available, these materials would be the best choice as the substrate for deep UV LEDs.

Free-standing GaN is a nearly perfect chemical, crystallographic, lattice-constant, and thermal-expansion match to LED heterostructures. Homoepitaxy would improve the quality and yield of LED production, and simplify the procedures for growth, fabrication, packaging, and thermal designs. Homoepitaxial LEDs have proven to be suitable for high-power and highly reliable operation, and thus more light can be delivered from a single package. Unfortunately the growth of bulk GaN is still in its infancy. Bulk GaN grown by HVPE has relatively poor crystalline quality, whereas crystals produced using solution growth techniques have slow rate and small size limitations. The lack of low-cost, large-size, and flawless GaN wafers remains an obstacle. This obstacle must be overcome before III–nitride LEDs grown on GaN substrates have an impact on the development of high-brightness and cost-efficient solid-state lighting sources.

Acknowledgments

The author would like to thank his collaborators S. F. LeBoeuf, J. Teetsov, J. W. Kretchmer, S. D. Arthur, D. W. Merfeld, and M. P. D'Evelyn at GE Global Research Center, and C. H. Yan and W. Wang at Lumei Optoelectronics.

References

1 S. Nakamura, S. J. Pearton, and G. Fasol, *The Blue Laser Diode*, Springer-Verlag, Heidelberg, Germany, **2000**.

2 M. A. Khan, M. Shatalov, H. P. Maruska, H. M. Wang, and E. Kuokstis, *Jpn. J. Appl. Phys.* **2005**, 44, 7191.

3 J. Y. Tsao, *IEEE Circuits and Devices* **2004**, 5/6, 28.

4 S. F. Chichibu, Y. Kawakami, and T. Sota, in *Introduction to Nitride Semiconductor Blue Lasers and Light Emitting Diodes*, ed. S. Nakamura and S. F. Chichibu, Taylor and Francis, New York, **2000**.

5 X. Xu, R. P. Vaudo, C. Lario, A. Saland, G. R. Brandes, and J. Chaudhuri, *J. Cryst. Growth* **2002**, 246, 223.

6 J. C. Rojo, G. A. Slack, K. Morgan, B. Raghothamachar, M. Dudley, and L. J. Schowalter, *J. Cryst. Growth* **2001**, 231, 317.

7 S. P. DenBaars and S. Keller, in *Gallium Nitride (II)*, ed. J. I. Pankove and T. D. Moustakas, Academic Press, San Diego, **1998**.

8 H. Amano, I. Akasaki, T. Kozawa, K. Hiramatsu, N. Sawak, K. Ikeda, and Y. Ishi, *J. Lumin.* **1988**, 40–41, 121.

9 S. Nakamura, N. Iwasa, M. Senoh, and T. Mukai, *Jpn. J. Appl. Phys.* **1992**, 31, 1258.

10 H. Amano, N. Sawaki, I. Akasaki, and Y. Toyoda, *Appl. Phys. Lett.* **1986**, 48, 353.

11 S. Nakamura, *Jpn. J. Appl. Phys.* **1991**, 30, L1705.

12 H. Kawakami, K. Sakurai, K. Tsubouchi, and N. Mikoshiba, *Jpn. J. Appl. Phys.* **1998**, 27, L161.

13 T. D. Moustakas, T. Lei, and R. J. Molnar, *Physica B* **1993**, 185, 36.

14 S. Nakamura, T. Mukai, and M. Senoh, *Appl. Phys. Lett.* **1994**, 64, 1687.

15 S. Nakamura, M. Senoh, N. Iwasa, S. Nagahama, T. Yamada, T. Mukai, *Jpn. J. Appl. Phys.* **1995**, 34, L1332.

16 D. Steigerwald, S. Rudaz, H. Liu, R. S. Kern, W. Götz, and R. Fletcher, *JOM* **1997**, 49, 18.

17 M. Koike, N. Shibata, H. Kato, and Y. Takahashi, *IEEE J. Sel. Top. Quantum Electron*, **2002**, 8, 271.

18 J. P. Zhang, A. Chitnis, V. Adivarahan, S. Wu, V. Mandavilli, R. Pachipulusu, M. Shatalov, G. Simin, J. W. Yang, and M. Asif Khan, *Appl. Phys. Lett.* **2002**, 81, 4910.

19 A. A. Allerman, M. H. Crawford, A. J. Fischer, K. H. A. Bogart, S. R. Lee, D. M. Follstaedt, P. P. Provencio, and D. D. Koleske, *J. Cryst. Growth*, **2004**, 272, 227.

20 M. Leszczynski, T. Suski, H. Teisseyre, P. Perlin, I. Grzegory, J. Jun, S. Porowski, and T. D. Moustakas, *J. Appl. Phys.* **1994**, 76, 4909.

21 L. T. Romano, C. G. Van de Walle, J. W. Ager III, W. Götz, and R. S. Kern, *J. Appl. Phys.* **2000**, 87, 7745.

22 J. Zhang, E. Kuokstis, Q. Fareed, H. Wang, J. Yang, G. Simin, M. Asif Khan, R. Gaska, and M. Shur, *Appl. Phys. Lett.* **2001**, 79, 925.

23 S. D. Lester, F. A. Ponce, M. G. Craford, and D. A. Steigerwald, *Appl. Phys. Lett.* **1995**, 66, 1249.

24 D. Kapolnek, X. H. Wu, B. Heying, S. Keller, U. K. Mishra, S. P. DenBaars, and S. J. Speck, *Appl. Phys. Lett.* **1995**, 67, 1541.

25 X. H. Wu, C. R. Elsass, A. Abare, M. Mack, S. Keller, P. M. Petroff, S. P. DenBaars, J. S. Speck, and S. J. Rosner, *Appl. Phys. Lett.* **1998**, 72, 692.

26 D. I. Florescu, S. M. Ting, J. C. Ramer, D. S. Lee, V. N. Merai, A. Parkeh, D. Lu,

E. A. Armour, and L. Chernyak, *Appl. Phys. Lett.* **2003**, 83, 33.

27 Z. Liliental-Weber, Y. Chen, S. Ruvimov, and J. Washburn, *Phys. Rev. Lett.* **1997**, 79, 2835.

28 A. H. Herzog, D. L. Keune, and M. G. Craford, *J. Appl. Phys.* **1972**, 43, 600.

29 S. D. Lester, F. A. Ponce, M. G. Craford, and D. A. Steigerwald, *Appl. Phys. Lett.* **1996**, 66, 1249.

30 T. Mukai and S. Nakamura, *Jpn. J. Appl. Phys.* **1999**, 38, 5735.

31 S. Chichibu, T. Aznhata, T. Sota, and S. Nakamura, *Appl. Phys. Lett.* **1996**, 69, 4188.

32 X. A. Cao, E. B. Stokes, P. Sandvik, N. Taskar, J. Kretchmer, and D. Walker, *Solid State Electron.* **2002**, 46, 1235.

33 H. Kim, S. J. Park, H. Hwang, and N. M. Park, *Appl. Phys. Lett.* **2002**, 81, 1326.

34 X. Guo, Y. L. Li, and E. F. Schubert, *Appl. Phys. Lett.* **2001**, 79, 1936.

35 J. K. Ho, C. S. Jong, C. C. Chiu, C. N. Huang, C. Y. Chen, and K. Shih, *Appl. Phys. Lett.* **1999**, 74, 1275.

36 J. K. Ho, C. S. Jong, C. C. Chiu, C. N. Huang, K., K. Shih, L. C. Chen, F. R. Chen, and J. J. Kai, *J. Appl. Phys.* **1999**, 86, 4491.

37 J. O. Song, J. S. Kwak, Y. Park, and T. Y. Seong, *Appl. Phys. Lett.* **2004**, 85, 6374.

38 J. O. Song, J. S. Kwak, Y. Park, and T. Y. Seong, *Appl. Phys. Lett.* **2005**, 86, 213505.

39 S. P. Jung, D. Ullery, C. H. Lin, H. P. Lee, J. H. Lim, D. K. Hwang, J. Y. Kim, E. J. Yang, and S. J. Park, *Appl. Phys. Lett.* **2005**, 87, 181107.

40 X. Guo and E. F. Schubert, *Appl. Phys. Lett.* **2001**, 78, 3337.

41 A. Chitnis, J. Sun, V. Mandavilli, R. Pachipulusu, S. Wu, M. Gaevski, V. Adivarahan, J. P. Zhang, M. A. Khan, A. Sarua, and M. Kuball, *Appl. Phys. Lett.* **2002**, 81, 3491.

42 S. M. Sze, *Physics of Semiconductor Devices*, 2nd edition, Wiley, New York, **1981**.

43 G. Franssen, E. Staszewska, R. Piotrzkowski, T. Suski, and P. Perlin, *J. Appl. Phys.* **2003**, 94, 6122.

44 D. J. Dumin and G. L. Pearson, *J. Appl. Phys.* **1965**, 36, 3418.

45 X. A. Cao, E. B. Stokes, P. Sandvik, S. F. LeBoeuf, J. Kretchmer, and D. Walker, *IEEE Electron Dev. Lett.* **2002**, 23, 535.

46 A. R. Riben and D. L. Feucht, *Solid-State Electron.* **1966**, 9, 1055.

47 S. R. Forrest, M. Didomernico, Jr., R. G. Smith, and H. J. Stocker, *Appl. Phys. Lett.* **1980**, 36, 580.

48 J. B. Fedison, T. P. Chow, H. Lu, and I. B. Bhat, *Appl. Phys. Lett.* **1998**, 72, 2841.

49 P. G. Eliseev, P. Perlin, J. Furioli, J. Mu, M. Banas, P. Sartori, J. Mu, and M. Osinski, *J. Electron. Mater.* **1997**, 26, 311.

50 Y. Narukawa, Y. Kawakami, S. Fujita, S. Fujita, and S. Nakamura, *Phys. Rev. B* **1997**, 55, 1938.

51 R. W. Martin, P. G. Middleton, K. P. O'Donnell, and W. Van der Stricht, *Appl. Phys. Lett.* **1999**, 74, 263.

52 H. C. Yang, P. F. Kuo, T. Y. Lin, Y. F. Chen, K. H. Chen, L. C. Chen, and J. I. Chyi, *Appl. Phys. Lett.* **2000**, 76, 3712.

53 L. Nistor, H. Bender, A. Vantomme, M. F. Wu, J. V. Lauduyt, K. P. O'Donnell, R. Martin, K. Jacobs, and I. Moerman, *Appl. Phys. Lett.* **2000**, 77, 507.

54 S. Chichibu, K. Wada, and S. Nakamura, *Appl. Phys. Lett.* **1997**, 71, 2346.

55 K. P. O'Donnell, R. W. Martin, and P. G. Middleton, *Phys. Rev. Lett.* **1999**, 82, 237.

56 C. H. Chen, L. Y. Huang, Y. F. Chen, H. X. Jiang, and J. Y. Lin, *Appl. Phys. Lett.* **2002**, 80, 1397.

57 G. Tamulaitis, K. Kazlauskas, S. Jursenas, A. Zukauskas, M. A. Khan, J. W. Yang, J. Zhang, G. Simin, M. S. Shur, and R. Gaska, *Appl. Phys. Lett.* **2000**, 77, 2136.

58 M. A. Khan, V. Adivarahan, J. P. Zhang, C. Chen, E. Kuokstis, A. Chitnis, M. Shatalov, J. W. Yang, and G. Simin, *Jpn. J. Appl. Phys.* **2001**, 40, L1308.

59 H. Hirayama, *J. Appl. Phys.* **2005**, 97, 091101.

60 F. G. McIntosh, K. S. Boutros, J. C. Roberts, S. M. Bedair, E. L. Piner, and N. A. El-Masry, *Appl. Phys. Lett.* **1996**, 68, 40.

61 F. Bernardini, V. Fiorentini, F. Della Sala, A. Di Carlo, and P. Lugli, *Phys. Rev. B* **1999**, 60, 8849.

62 T. Deguchi, A. Shikanai, K. Torii, T. Sota, S. Chichibu, and S. Nakamura, *Appl. Phys. Lett.* **1998**, 72, 3329.

63 T. Takeuchi, S. Sota, M. Katsuragawa, M. Komori, H. Takeuchi, H. Amano, and I. Akasaki, *Jpn. J. Appl. Phys.* **1997**, 36, L382.

64 Y. D. Qi, H. Liang, D. Wang, Z. D. Lu, W. Tang, and K. M. Lau, *Appl. Phys. Lett.* **2005**, 86, 101903.

65 Y. H. Cho, J. J. Song, S. Keller, M. S. Minsky, E. Hu, U. K. Mishra, and S. P. DenBaars, *Appl. Phys. Lett.* **1998**, 73, 1128.

66 A. E. Yunovich and S. S. Mamakin, *Mat. Res. Soc. Symp. Proc.* **2002**, 722, K2.4.1.

67 Y. H. Cho, G. H. Gainer, A. J. Fischer, J. J. Song, S. Keller, U. K. Mishra, and S. P. DenBaars, *Appl. Phys. Lett.* **1998**, 73, 1370.

68 W. Shan, T. J. Schmidt, X. H. Yang, S. J. Hwang, J. J. Song, and B. Goldenberg, *Appl. Phys. Lett.* **1995**, 66, 985.

69 I. L. Krestnikov, N. N. Ledentsov, A. Hoffmann, D. Bimberg, A. V. Sakharov, W. V. Lundin, A. F. Tsatsulnikov, A. S. Usikov, Z. I. Alferov, Y. G. Musikhin, and D. Gerthsen, *Phys. Rev. B* **2002**, 66, 155310.

70 M. Smith, G. D. Chen, J. Y. Lin, H. X. Jiang, M. Asif Khan, and Q. Chen, *Appl. Phys. Lett.* **1996**, 69, 2837.

71 B. Monemar, *Phys. Rev. B* **1974**, 10, 676.

72 S. Nakamura, *JSAP International*, **2000**, 1, 5.

73 M. Khizar, Z. Y. Fan, K. H. Kim, J. Y. Lin, and H. X. Jiang, *Appl. Phys. Lett.* **2005**, 85, 173504.

74 K. Mayes, A. Yasan, R. McClintock, D. Shiell, S. R. Darvish, P. Kung, and M. Razeghi, *Appl. Phys. Lett.* **2004**, 84, 1046.

75 A. J. Fischer, A. A. Allerman, M. H. Crawford, K. H. A. Bogart, S. R. Lee, R. J. Kaplar, W. W. Chow, S. R. Kurtz, K. W. Fullmer, and J. J. Figiel, *Appl. Phys. Lett.* **2004**, 84, 3394.

76 J. P. Zhang, S. Wu, S. Rai, V. Mandavilli, V. Adivarahan, A. Chitnis, M. Shatalov, and M. A. Khan, *Appl. Phys. Lett.* **2003**, 83, 3456.

77 A. R. Riben and D. L. Feucht, *Solid-State Electron.* **1966**, 9, 1055.

78 X. A. Cao, S. F. Leboeuf, L. R. Rowland, and H. Liu, *J. Electron. Mater.* **2003**, 32, 316.

79 J. J. Wierer, D. A. Steigerwald, M. R. Krames, J. J. O'Shea, M. J. Ludowise, G. Christenson, Y.-C. Shen, C. Lowery, P. S. Martin, S. Subramanya, W. Gotz, N. F. Gardner, R. S. Kern, and S. A. Stockman, *Appl. Phys. Lett.* **2001**, 78, 3379.

80 J. C. Bhat, A. Kim, D. Collins, R. Fletcher, R. Khare, and S. Rudaz, *ISCS*, Tokyo, Japan, October **2001**.

81 T. Fujii, Y. Gao, R. Sharma, E. L. Hu, S. P. DenBaars, and S. Nakamura, *Appl. Phys. Lett.* **2004**, 84, 855.

82 C. Huh, K. S. Lee, E. J. Kang, and S. J. Park, *J. Appl. Phys.* **2003**, 93, 9383.

83 K. Tadatomo, H. Okagawa, Y. Ohuchi, T. Tsunekawa, Y. Imada, M. Kato, and T. Taguchi, *Jpn. J. Appl. Phys.* **2001**, 40, L583.

84 M. Yamada, T. Mitani, Y. Narukawa, S. Shioji, I. Niki, S. Sonobe, K. Deguchi, M. Sano, and T. Mukai, *Jpn. J. Appl. Phys.* **2002**, 41, L1431.

85 M. Khizar, Z. Y. Fan, K. H. Kim, J. Y. Lin, and H. X. Jiang, *Appl. Phys. Lett.* **2005**, 86, 173504.

86 T. N. Oder, K. H. Kim, J. Y. Lin, and H. X. Jiang, *Appl. Phys. Lett.* **2004**, 84, 466.

87 J. J. Wierer, M. R. Krames, J. E. Epler, N. F. Gardner, M. G. Craford, J. R. Wendi, J. A. Simmons, and M. M. Sigalas, *Appl. Phys. Lett.* **2004**, 84, 3885.

88 K. Okamoto, I. Niki, A. Shvartser, Y. Narukawa, T. Mukai, and A. Scherer, *Nat. Mater.* **2004**, 3, 601.

89 J. Edmond and J. Lagaly, *JOM* **1997**, 9, 24.

90 V. Harle, B. Hahn, H. J. Lugaauer, G. Bruderl, D. Eisert, U. Strauss, A. Lell, and N. Hiller, *Compound Semiconductors* **2000**, 6, 81.

91 F. A. Ponce, B. S. Krusor, J. S. Major, Jr., W. E. Plano, and D. F. Welch, *Appl. Phys. Lett.* **1995**, 67, 410.

92 T. W. Weeks, Jr., M. D. Bremser, K. S. Ailey, E. Carlson, W. G. Perry, and R. F. Davis, *Appl. Phys. Lett.* **1995**, 67, 401.

93 S. Tanaka, S. Iwai, and Y. Aoyagi, *J. Cryst. Growth* **1997**, 170, 329.

94 P. Waltereit, O. Brandt, A. Trampert, M. Ramsteiner, M. Reiche, M. Qi, and K. H. Ploog, *Appl. Phys. Lett.* **1999**, 74, 3660.

95 M. R. Krames, M. Ochiai-Holcomb, G. E. Hofler, C. Carter-Coman, E. I. Chen, I. H. Tan, P. Grillot, N. F. Gardner, H. C. Chui, J.-W. Huang, S. A. Stockman, F. A. Kish, and M. G. Craford, *Appl. Phys. Lett.* **1999**, 75, 2365.

96 J. Edmond, D. Emerson, M. Bergmann, K. Haberer, and C. Hussell, *ICSCRM*, Pittsburgh, PA, **2005**.

97 S. Guha and N. A. Bojarczuk, *Appl. Phys. Lett.* **1998**, 73, 1487.

98 G. Kipshidze, V. Kuryatkov, B. Borisov, M. Holtz, S. Nikishin, and H. Temkin, *Appl. Phys. Lett.* **2002**, 80, 3682.

99 C. A. Tran, A. Osinski, R. F. Karlicek, Jr., and I. Berishev, *Appl. Phys. Lett.* **1999**, 75, 1494.

100 A. Dadgar, M. Poschenrieder, J. Bläsing, K. Fehse, A. Diez, and A. Krost, *Appl. Phys. Lett.* **2002**, 80, 3670.

101 H. Ishikawa, B. Zhang, K. Asano, T. Egawa, and T. Jimbo, *J. Cryst. Growth* **2004**, 272, 322.

102 B. Zhang, T. Egawa, H. Ishikawa, Y. Liu, and T. Jimbo, *Appl. Phys. Lett.* **2005**, 86, 071113.

103 J. Li, J. Y. Lin, and H. X. Jiang, MRS Fall Meeting, Boston, MA, **2004**.

104 H. P. D. Schenk, G. D. Kipshidze, V. B. Lebedev, S. Shokhovets, R. Goldhahn, J. Kräußlich, A. Fissel, and W. Richter, *J. Cryst. Growth* **1999**, 201–202, 359.

105 S. A. Nikishin, N. N. Faleev, A. S. Zubrilov, V. G. Antipov, and H. Temkin, *Appl. Phys. Lett.* **2000**, 76, 3028.

106 H. Marchand, L. Zhao, N. Zhang, B. Moran, R. Coffie, U. K. Mishra, J. S. Speck, S. P. DenBaars, and J. A. Freitas, *J. Appl. Phys.* **2001**, 89, 7846.

107 M.-H. Kim, Y.-C. Bang, N.-M. Park, C.-J. Choi, T.-Y. Seong, and S.-J. Park, *Appl. Phys. Lett.* **2001**, 78, 2858.

108 E. Feltin, B. Beaumont, M. Laügt, P. de Mierry, P. Vennéguès, H. Lahrèche, M. Leroux, and P. Gibart, *Appl. Phys. Lett.* **2001**, 79, 3230.

109 W. S. Wong, T. Sands, N. W. Cheung, M. Kneissl, D. P. Bour, P. Mei, L. T. Romano, and N. M. Johnson, *Appl. Phys. Lett.* **2000**, 77, 2822.

110 B. S. Tan, S. Yuan, and X. J. K. Kang, *Appl. Phys. Lett.* **2004**, 84, 2757.

111 S. Porowski and I. Grzegory, in *GaN and Related Materials*, ed. S. J. Pearton, Gordon and Breach, New York, p. 295, **1997**.

112 S. Porowski, *J. Cryst. Growth* **1996**, 166, 583.

113 S. Sakai, S. Sato, T. Sugahara, Y. Naoi, S. Kurai, K. Yamashita, S. Tottori, M. Hao, K. Wada, and K. Nishino, *Mater. Sci. Forum* **1998**, 264/268, 1107.

114 T. Inoue, Y. Seki, O. Oda, S. Kurai, Y. Yamada, and T. Taguchi, *Phys. Stat. Sol. (b)* **2001**, 223, 15.

115 C. M. Balkas, Z. Sitar, L. Bergman, I. K. Shmagin, J. F. Muth, R. Kolbas, R. J. Nemanich, and R. F. Davis, *J. Cryst. Growth* **2000**, 208, 100.

116 Y. Kumagai, H. Murakami, A. Koukitu, K. Takemoto, and H. Seki, *Jpn. J. Appl. Phys.* **2000**, 39, L703.

117 K. Lee and K. Auh, *MRS Internet J. Nitride Semicond. Res.* **2001**, 6, 4.

118 D. Gogova, A. Kasic, H. Larsson, C. Hemmingsson, B. Monemar, F. Tuomisto, K. Saarinen, L. Dobos, B. Pecz, P. Gibart, and B. Beaumont, *J. Appl. Phys.* **2004**, 96, 799.

119 X. Xu, R. P. Vaudo, C. Loria, A. Salant, G. R. Brandes, and J. Chaudhuri, *J. Cryst. Growth* **2002**, 246, 223.

120 X. Xu, R. P. Vaudo, and G. R. Brandes, *Opt. Mater.* **2003**, 23, 1.

121 X. Xu, R. P. Vaudo, J. Flynn, J. Dion, and G. R. Brandes, *Phys. Stat. Sol. (a)* **2005**, 202, 727.

122 C. R. Miskys, M. K. Kelly, O. Ambacher, G. Martinez-Criado, and M. Stutzmann, *Appl. Phys. Lett.* **2000**, 77, 1858.

123 M. Schauler, F. Eberhard, C. Kirchner, V. Schwegler, A. Pelzmann, M. Kamp, K. J. Ebeling, F. Bertram, T. Riemann, J. Christen, P. Prystawko, M. Leszczynski, I. Grzegory, and S. Porowski, *Appl. Phys. Lett.* **1999**, 74, 1123.

124 X. A. Cao, H. Lu, S. F. LeBoeuf, C. Cowen, S. D. Arthur, and W. Wang, *Appl. Phys. Lett.* **2005**, 87, 053503.

125 I. Arslan and N. D. Browning, *Phys. Rev. Lett.* **2003**, 91, 165501–1.

126 A. Pelzmann, C. Kirchner, M. Mayer, V. Schwegler, M. Schauler, M. Kamp, K. J. Ebeling, I. Grzegory, M. Leszczynski, G. Nowak, and S. Porowski, *J. Cryst. Growth* **1998**, 189–190, 167.

127 G. Franssen, E. Litwin-Staszewska, R. Piotrzkowski, T. Suski, and P. Perlin, *J. Appl. Phys.* **2003**, 94, 6122.

128 X. A. Cao, J. Teetsov, F. Shahedipour-Sandvik, and S. D. Arthur, *J. Cryst. Growth* **2004**, 264, 172.

129 Y. Chen, T. Takeuchi, H. Amano, I. Akasaki, N. Yamada, Y. Kaneko, and S. Y. Wang, *Appl. Phys. Lett.* **1998**, 72, 710.

130 E. J. Miller, D. M. Schaadt, E. T. Yu, C. Poblenz, C. Elsass, and J. S. Speck, *J. Appl. Phys.* **2002**, 91, 9821.

131 J. W. P. Hsu, M. J. Manfra, D. V. Lang, S. Richter, S. N. G. Chu, A. M. Sergent, R. N. Kleiman, L. N. Pfeiffer, and R. J. Molnar, *Appl. Phys. Lett.* **2001**, 78, 1685.

132 A. A. Pomarico, D. Huang, J. Dickinson, A. A. Baski, R. Cingolani, H. Morkoc, and R. Molnar, *Appl. Phys. Lett.* **2003**, 82, 1890.

133 T. Nishida, H. Saito, and N. Kobayashi, *Appl. Phys. Lett.* **2001**, 79, 711.

134 A. Yasan, R. McClintock, K. Mayes, S. R. Darvish, H. Zhang, P. Kung, M. Razeghi, S. K. Lee, and J. Y. Han, *Appl. Phys. Lett.* **2002**, 81, 2151.

135 G. A. Garrett, C. J. Collins, A. V. Sampath, H. Shen, M. Wraback, S. F. LeBoeuf, J. Flynn, and G. Brandes, *Phys. Stat. Sol. (c)* **2005**, 2, 2332.

136 M. G. Craford, *Proc. SPIE* **2002**, 4776, 1.

137 L. Sugiura, *Appl. Phys. Lett.* **1997**, 70, 1317.

138 M. Kneissl, D. P. Bour, L. Romano, C. G. Van de Walle, J. E. Northrup, W. S. Wong, D. W. Treat, M. Teepe, T. Schmidt, and N. M. Johnson, *Appl. Phys. Lett.* **2000**, 77, 1931.

139 S. Nakamura, M. Senoh, S. Nagahama, N. Iwasa, T. Yamada, T. Matsushita, H. Kiyoku, and Y. Sugimoto, *Appl. Phys. Lett.* **1998**, 72, 211.

140 S. Nakamura, M. Senoh, S. Nagahama, N. Iwasa, T. Matsushita, and T. Mukai, *Appl. Phys. Lett.* **2000**, 76, 22.

141 S. Nakamura, M. Senoh, S. Nagahama, N. Iwasa, T. Yamada, T. Matsushita, H. Kiyoku, Y. Sugimoto, T. Kozaki, H. Umemoto, M. Sano, and K. Chocho, *Appl. Phys. Lett.* **1998**, 73, 832.

142 G. A. Slack and T. F. McNelly, *J. Cryst. Growth* **1976**, 34, 163.

143 J. C. Rojo, G. A. Slack, K. Morgan, B. Raghothamachar, M. Dudley, and L. J. Schowalter, *J. Cryst. Growth* **2001**, 231, 317.

144 L. J. Schowalter, G. A. Slack, J. B. Whitlock, K. Morgan, S. B. Schujman, B. Raghothamachar, M. Dudley, and K. R. Evans, *Phys. Stat. Sol. (c)* **2003**, 0, 1997.

145 J. C. Rojo, L. J. Schowalter, R. Gaska, M. S. Shur, M. A. Khan, J. Yang, and D. D. Koleske, *J. Cryst. Growth* **2002**, 240, 508.

146 S. A. Nikishin, B. A. Borisov, A. Chandolu, V. V. Kuryatkov, H. Temkin, M. Holtz, E. N. Mokhov, Y. Makarov, and H. Helava, *Appl. Phys. Lett.* **2004**, 85, 4355.

147 R. Gaska, C. Chen, J. Yang, E. Kuokstis, A. Khan, G. Tamulaitis, I. Yilmaz, M. S. Shur, J. C. Rojo, and L. J. Schowalter, *Appl. Phys. Lett.* **2002**, 81, 4658.

148 T. Nishida, T. Makimoto, H. Saito, and T. Ban, *Appl. Phys. Lett.* **2004**, 84, 1002.

149 G. Namkoong, S. Burnham, K. K. Lee, E. Trybus, W. A. Doolittle, M. Losurdo, P. Capezzuto, G. Bruno, B. Nemeth, and J. Nause, *Appl. Phys. Lett.* **2005**, 87, 184104.

150 W. A. Doolittle and A. S. Brown, *Solid State Electron.* **2000**, 44, 229.

151 S. Kang, W. A. Doolittle, A. S. Brown, and S. R. Stock, *Appl. Phys. Lett.* **1999**, 74, 3380.

152 S. Kamiyama, M. Iwaya, H. Amano, and I. Akasaki, *Mater. Res. Soc. Symp. Proc.* **2005**, 831.

153 A. Chakraborty, B. A. Haskell, S. Keller, J. S. Speck, S. P. DenBaars, S. Nakamura, and U. K. Mishra, *Appl. Phys. Lett.* **2004**, 85, 5143.

154 A. Chitnis, C. Chen, V. Adivarahan, M. Shatalov, E. Kuokstis, V. Mandavilli, J. Yang, and M. A. Khan, *Appl. Phys. Lett.* **2004**, 84, 3663.

155 C. Chen, V. Adivarahan, J. Yang, M. Shatalov, E. Kuokstis, and M. A. Khan, *Jpn. J. Appl. Phys.* **2003**, 42, L1039.

156 A. Chakraborty, B. A. Haskell, S. Keller, J. S. Speck, S. P. Denbaars, S. Nakamura, and U. K. Mishra, *Jpn. J. Appl. Phys.* **2005**, 44, L173.

157 N. F. Gardner, J. C. Kim, J. J. Wierer, Y. C. Shen, and M. R. Krames, *Appl. Phys. Lett.* **2005**, 86, 111101.

158 M. D. Craven, S. H. Lim, F. Wu, J. S. Speck, and S. P. DenBaars, *Appl. Phys. Lett.* **2002**, 81, 469.

159 W. H. Sun, J. W. Yang, C. Q. Chen, J. P. Zhang, M. E. Gaevski, E. Kuokstis, V. Adivarahan, H. M. Wang, Z. Gong, M. Su, and M. A. Khan, *Appl. Phys. Lett.* **2003**, 83, 2599.

160 K. Domen, K. Horino, A. Kuramata, and T. Tanahashi, *Appl. Phys. Lett.* **1997**, 71, 1996.

161 A. Niwa, T. Ohtoshi, and T. Kuroda, *Jpn. J. Appl. Phys.* **1996**, 35, L599.

162 H. Masui, A. Chakraborty, B. A. Haskell, U. K. Mishra, J. S. Speck, S. Nakamura, and S. P. DenBaars, *Jpn. J. Appl. Phys.* **2005**, 44, L1329.

2
III–Nitride Microcavity Light Emitters

Hongxing Jiang and Jingyu Lin

2.1
Introduction

Devices based on III–nitrides offer a great potential for applications such as ultra-violet (UV) and blue lasers, solar-blind UV detectors, and high-power electronics. Researchers in this field have made extremely rapid progress toward materials growth as well as device fabrication [1–6]. The successes of super-bright blue light-emitting diodes (LEDs) and UV/blue laser diodes (LDs) based on the III–nitride system are a clear indication of the great potential of this material system. The recent success of the III–nitride edge emitters and detectors is encouraging for the study of microscale photonic structures and devices [7]. These microphotonic devices range from arrays of microemitters, detectors, and waveguides to optical switches and photonic crystals. New physical phenomena and properties begin to dominate as the device size scale approaches the wavelength of light, including modified spontaneous emission, enhanced quantum efficiency, and lasing in microcavities, all of which warrant fundamental investigations [8–12].

The micro- and nanophotonic technologies, like the integration of Si transistors in the 1960s, are expected eventually to provide the ability to integrate arrays of thousands of cheaply fabricated optical circuit elements such as waveguides, switches, resonators, etc. on a single chip. The physics of microphotonic structures and devices has been investigated and much progress has been made in the past decade. Although many of the ideas and potentials of these devices were identified long ago, it is only recently that the transition from basic research to practical device components has been made for microscale photonic devices due to various technological advances. Together with their potential for optical circuit element integration, these microphotonic devices open many important applications such as optical communications, signal and image processing, optical interconnects, computing, enhanced energy conversion and storage, and chemical, biohazard substance, and disease detection.

III–nitride optoelectronic devices offer special benefits including UV/blue emission (allowing higher optical storage density and resolution as well as the ability

Wide Bandgap Light Emitting Materials and Devices. Edited by G. F. Neumark, I. L. Kuskovsky, and H. Jiang
Copyright © 2007 WILEY-VCH Verlag GmbH & Co. KGaA, Weinheim
ISBN: 978-3-527-40331-8

for chemical and biohazard substance detection), the ability to operate at very high temperatures and power levels due to their mechanical hardness and larger band-gaps, high speed due to the intrinsically rapid radiative recombination rates, and large band offset of 2.7 eV or 5.4 eV for GaN/AlN or InN/AlN heterostructures allowing novel quantum-well (QW) device designs, and high emission efficiencies. These unique properties may allow the creation of microscale opto-electronic and photonic devices with unprecedented properties and functions. However, these unique properties also make the device physics and architectures of the nitrides different and much more challenging than other conventional semiconductors. For example, most of the nitride emitter structures are grown on sapphire substrates, which are insulating and cause extra difficulties in Ohmic contact fabrication. Additionally, the conductivity of p-type materials is much lower for the nitrides, which makes hole injection very difficult and requires innovative methods to improve the p-type conductivity as well as the current spreading.

In this chapter, we summarize recent progress in III–nitride microscale structures and light emitters. In Section 2.2, we discuss the fabrication of III–nitride microstructures including microdisks, microrings, micropyramids, and submicron waveguides. In Section 2.3, we summarize optical studies, primarily probed by photoluminescence (PL) emission spectroscopy, carried out on III–nitride microstructures. Section 2.4 presents the fabrication and characterization of current-injected microscale emitters. In Section 2.5, we show examples of approaches for fabricating practical device components. We first discuss the application of micro-LEDs for boosting the output power of LEDs. By interconnecting hundreds of III–nitride micro-LEDs (size on the order of 10 μm in diameter) and fitting them into the same device area as taken up by a conventional LED, an enhancement of up to 40% in LED output power was demonstrated. We then discuss the fabrication and characteristics of micro-LED arrays with addressable individual pixels for microdisplays and other applications. By controlling each micro-LED individually in a micro-LED array, the first semiconductor microdisplay was demonstrated using nitride materials. Applications of this novel device for optical communication and entertainment are discussed. The characteristics of this novel semiconductor display are also compared with that of organic LED as well as liquid-crystal microdisplays.

2.2
III–Nitride Microstructure Fabrication

2.2.1
Microstructure Fabrication by Photolithography Patterning and Plasma Dry Etching

III–nitride microstructures, including microdisks, microrings, submicron wave-guides, microscale prisms and pyramids, have been fabricated by several methods. In the early phase of our studies, ion-beam etching was used to prepare III–nitride microstructures [13–15]. More recently, the authors' laboratory has employed the inductively coupled plasma (ICP) etching technique to pattern the III–nitride microstructures fabricated from multiple-quantum-well (MQW) structures, which were grown on (0001) sapphire substrates by metal–organic chemical vapor deposition (MOCVD). ICP etching has been shown to be very effective for GaN etching with high etch rate but minimal ion damage [16]. Photolithography was used to pattern arrays of microstructures with different dimensions and separation spacing. The samples were etched by ICP into the sapphire substrate so that no III–nitride material is present between microstructures. Figure 2.1(a) shows a schematic diagram of the AlGaN/GaN MQW sample containing circular photoresist masks and Fig. 2.1(b) shows the AlGaN/GaN MQW microdisks. Arrays of microdisks and microrings were fabricated from GaN/AlGaN and InGaN/GaN MQWs with dimension from 5 to 20 µm [12–15, 17]. Figure 2.1(c) shows the scanning electron microscopy (SEM) image of a representative III–nitride QW microdisk.

In another case, the disks were fabricated from a controlled two-stage etching process from InGaN/GaN QWs grown on a thick AlN buffer layer by metal–organic molecular-beam epitaxy (MOMBE) [18]. After lithographically patterning circular photoresist masks, microcylinders are formed by electron cyclotron

(a) GaN/ALGaN MQWs

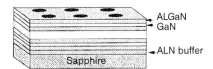

(b) GaN/ALGaN Micro-disk

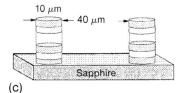

(c)

Fig. 2.1 (a) Schematic of AlGaN/GaN MQW with circular masks. (b) Schematic of AlGaN/GaN MQW microdisks. (c) SEM image of a microdisk fabricated from AlGaN/GaN MQWs. After Refs. [13–15]

Fig. 2.2 SEM images of a microdisk array (a) after ECR etching and (b) after selective wet etch undercut of the AlN layer, fabricated from MOMBE. After Ref. [18].

resonance (ECR) plasma etching at 170 °C in a $Cl_2/CH_4/H_2/Ar$ discharge at 1 mTorr pressure and a microwave power of 850 W. Additional radiofrequency (RF) power of 150 W was applied to the sample position to increase the ion energy to about 175 eV and thereby improve etch anisotropy. The AlN is then selectively wet etched in AZ 400K developer solution for about 30 min at 85 °C to produce an undercut and leave the InGaN/GaN disk supported on an AlN pedestal. Figure 2.2 shows an SEM image of a microdisk array after ECR etching and after selective wet etch undercut of the AlN, fabricated from MOMBE.

GaN microdisk structures have also been obtained by chemically assisted ion-beam etching (CAIBE) together with dopant-selective wet etching from GaN p–n homostructures with 1 μm n-type ($n \sim 1 \times 10^{18}\,cm^3$) layer and a 0.4 μm p-type GaN ($p \sim 3 \times 10^{17}\,cm^3$) cap layer grown on an SiC substrate [19]. Circular mesas with 4 μm diameter were first dry etched using CAIBE through the top p-type GaN layer and into the underlying n-type GaN layer. The sample was then wet etched by the photoelectrochemical (PEC) process to form the "mushroom-like" structure shown in Fig. 2.3. The top p-type GaN layer was not etched by the PEC process, but the n-type GaN layer below was selectively removed, resulting in the observed under-cutting of the top p-type layer.

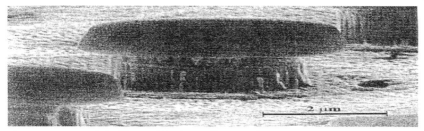

Fig. 2.3 GaN microdisks fabricated by chemically assisted ion-beam etching (CAIBE) together with dopant-selective wet etching from GaN p–n homostructures. After Ref. [19].

2.2.2
Microscale Pyramids and Prisms Fabricated by Selective Epitaxial Overgrowth

It has been shown that self-organized GaN microcavities produced by selective epitaxial overgrowth were either hexagonal prisms or hexagonal pyramids due to the nature of the crystal structures of GaN. GaN microprisms and micropyramids have been obtained through self-organization by selective epitaxial overgrowth using patterned GaN substrates [20–26]. Selective epitaxial overgrowth has the advantage that photonic structures can be fabricated without process damage. The generic procedure of self-organized III–nitride microstructure array preparation consists of the following basic steps: (a) first deposit a GaN epilayer on a sapphire substrate using a low-temperature buffer layer; (b) a thin capping film of SiO_2 or Si_xN_{1-x} was grown to act as a mask for further selective growth; (c) coating photoresist; (d) photolithography patterning; (e) ICP dry etching will then be carried out on SiO_2 or Si_xN_{1-x} layer to open circular window arrays with different sizes; and finally (f) these would then be followed by the overgrowth of self-organized III–nitride photonic structures and the removal of the mask layer by wet etching.

Figure 2.4(a) shows a schematic diagram of GaN micropyramids fabricated by selective overgrowth on GaN/AlN/silicon or GaN/AlN/sapphire substrates. Figure 2.4(b) shows an SEM image of an array of GaN hexagonal micropyramids obtained by selective epitaxial overgrowth. As can be seen from Fig. 2.4, the six facets of each pyramid formed by self-organization are extremely smooth since their formation does not require any patterning processes such as etching. These GaN pyramids formed a 2D array. The length of each side of the base of the self-organized pyramids was about 7 μm and the height of the pyramid was about 14 μm. Selectively grown GaN pyramids were also utilized to fabricate InGaN quantum dots (QDs), in which case InGaN QD structures of about 30 nm were formed on the top of the hexagonal pyramids. Intense photoluminescence was observed from the QD region [26].

GaN hexagonal microprisms of 5–16 μm diameter, with smooth vertical facets and no ridge growth, were fabricated on sapphire substrate by selective MOCVD growth [22]. An SEM image of GaN selectively grown on a hexagonal or circular

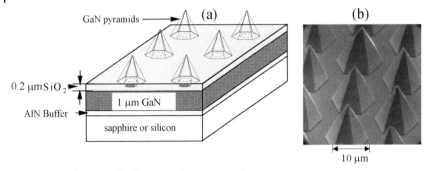

Fig. 2.4 (a) Schematic of self-organized GaN pyramid microcavities. (b) SEM image of an array of GaN hexagonal microscale pyramids obtained by selective MOCVD epitaxial overgrowth through self-organization. After Ref. [25].

GaN Microprisms

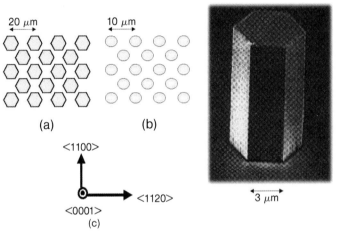

Fig. 2.5 SEM image of a GaN hexagonal microprism obtained by selective MOCVD epitaxial overgrowth through self-organization. After Ref. [22].

pattern of SiO$_2$ with a diameter of 5 µm is shown in Fig. 2.5. An 8 µm high GaN hexagonal microprism with a smooth top (0001) surface and vertical facets formed by self-organization, in spite of the circular opening. It was observed that the selection of the mask-patterning direction is an important factor, due to the 30° rotation of the crystallographic orientation between the GaN layer and the sapphire substrate. Dislocations of comparably high density ($\sim$10^9 cm^{-2}), which can be seen as lines vertical to the substrate, exist near the bottom of the GaN hexagonal microprism. But these defects mostly disappear near the top of the GaN hexagonal microprism, indicating that the crystallinity of the GaN film improves with increasing film thickness. The microprism provides strong optical confinement because the lasing light can achieve complete reflectance in the inscribed hexagonal optical path.

2.2.3
Submicron Waveguides

Waveguides are an important component in integrated photonic circuits. For III–nitride materials, there are two main challenges in the fabrication of submicron structures and devices. The first is the ability to define the structures through an appropriate lithography technique. Using standard photolithography it is difficult to achieve structures of dimensions less than 1 μm in III–nitrides due to limitations with photoresists, the wavelength of the light used, and the need for appropriate masks. The second challenge is the capability to transfer the designed submicron and nanoscale patterns to the samples. Because the III–nitrides are hard materials, the usual method to achieve this is by high-density plasma or ion etching. This requires a resist or a lift-off material that will withstand the plasma or ion etching as well as maintain the reduced dimensions of the structures. Submicron waveguide patterns based on AlGaN/GaN MQW have been successfully fabricated [27–29]. Success in the fabrication of these submicron structures opens many possibilities for nitride research for submicron and nanoscale device applications.

To provide lateral confinement, the waveguides were defined by electron-beam lithography technique using the Nanometer Pattern Generation System (NPGS). Following the standard degrease clean, two drops of negative resist were spun at 8000 rpm for 30 s to yield an estimated resist thickness of about 1.2 μm. A pre-exposure oven bake was carried out at 120 °C for 30 min. The electron beam used during the pattern writing was accelerated at 35 kV with a probe current of 5 pA and an areal dose of 5 μC cm^{-2} using a scanning electron microscopy system (model LEO 440). A post-exposure hot-plate bake was carried out at 105 °C for 5 min followed by developing for 90 s in dilute aqueous alkali solution (AZ 400K). The defined patterns were transferred to the sample by ICP etching. Dry etching was done at 300 W for 1 min, which resulted in an etch depth of about 0.75 μm. The resist used not only maintained submicron definition of patterns, but also held through the etching process, and the resulting waveguide patterns were quite satisfactory. Arrays of waveguide patterns occupying a 500 μm × 500 μm field with a center-to-center separation of 1 mm were defined. The waveguides were patterned with widths varying from 0.5 to 2.0 μm and orientations varying from −30° to 60° relative to the *a*-axis of GaN. Each waveguide was 500 μm in length and spaced about 15 μm from each other. For comparison purposes, a region of size 500 μm × 500 μm was defined and left unetched in the sample.

Figure 2.6(a) shows a schematic diagram of a waveguide structure, which is fabricated from AlGaN/GaN MQWs grown on sapphire by MOCVD. The MQW includes 30 periods of Al$_{0.2}$Ga$_{0.8}$N(50 Å)/GaN(12 Å) and followed by a 200 Å thick

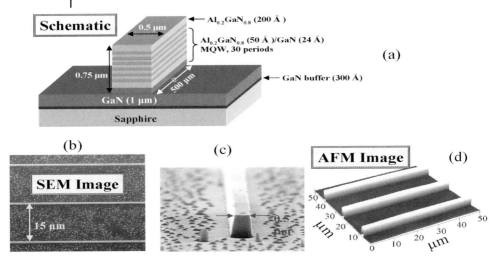

Fig. 2.6 (a) Schematic diagram showing the AlGaN/GaN QW waveguide structure. (b, c) Scanning electron microscope (SEM) images of the waveguide sample. The spacing between the waveguides is 15 μm and the width of the waveguides is 0.5 μm. (d) Atomic force microscope (AFM) image of the waveguides. After Ref. [27].

$Al_{0.2}Ga_{0.8}N$. Figures 2.6(b) and (c) are the SEM images of the waveguides. The width of the waveguides was about 0.5 μm as targeted. The etched surfaces of these structures were found to be quite smooth and the uniformity along the length of the waveguides was excellent. From the SEM image, there was no definite pattern found in the etch pits visible in the image. Figure 2.6(d) shows an atomic force microscope (AFM) image of the sample. From such images, we did not notice any preferred pattern in the edges of the etched steps of the waveguides for the different orientations.

2.3
Optical Studies of III–Nitride Microstructures

Photoluminescence (PL) emission spectroscopy was employed as a primary tool to study the optical properties of III–nitride microstructures. For PL studies, the authors' laboratory employs ultraviolet (UV) excitation pulses with pulse width of about 7 ps at a repetition rate of 9.5 MHz, which were provided by a picosecond UV laser system consisting of an yttrium–aluminum–garnet (YAG) laser (Coherent Antares 76) with a frequency doubler that pumps a cavity-dumped dye laser (Coherent 702–2CD) with Rhodamine 6G dye solution. A second frequency doubler was placed after the dye laser to provide a tunable photon energy up to 4.5 eV [30]. The laser output after the second doubler has an average power of about 30 mW and a spectral width of about 0.2 meV. The excitation intensity was controlled by a set of UV neutral density filters. Two detection systems were used to record the time-integrated and time-resolved PL emission signal. A single photon counting detection system with a microchannel-plate photomultiplier tube (MCP-PMT) has a time resolution of about 25 ps. A streak camera (Hamamatsu-C5680) detector has a time resolution of about 2 ps. To study the behavior of the optical resonance modes, a UV-transmitting objective was used in a confocal geometry to optically pump a single microstructure normal to the sample surface and to collect the light emission in the direction of the surface normal. Focused beam spot diameters as small as 2 µm could be achieved with the UV objective lens [13–15, 17].

2.3.1
Microdisks

Figure 2.7(a) shows continuous-wave (CW) spectra for the GaN/AlGaN MQWs at 10 K after and prior to the formation of microdisks. The generic structure of the GaN/AlGaN MQW samples used to fabricate microdisks consisted of a 30 nm AlN buffer layer followed by a ten-period 50 Å/50 Å GaN/Al$_x$Ga$_{1-x}$N ($x \sim 0.07$) MQW

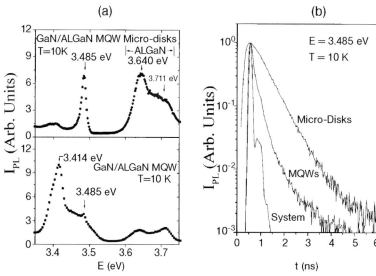

Fig. 2.7 (a) CW PL spectra at 10 K from the MQW microdisk structure (top) and the MQW structure (bottom). The spectra were obtained under similar excitation intensities and pump geometries. (b) Temporal response of the intrinsic exciton recombination measured at $E = 3.485$ eV and $T = 10$ K. Decays are shown for the sample prior to and after the formation of the microdisks. System response in this case is about 25 ps and is shown as indicated. After Ref. [13].

and a 200 Å AlN cap layer. All layers were grown nominally undoped. Comparison of the spectra in Fig. 2.7(a) reveals a large enhancement of the optical transitions at 3.485 and 3.640 eV in the MQW microdisks relative to the dominant transition (3.414 eV) prior to patterning. The obvious difference between the CW spectra seen here is very interesting because it suggests that the carrier dynamics and quantum efficiency within the MQW structure change upon the formation of the microdisks [13–15].

In order to elucidate the nature of the observed excitonic transition quantum efficiency enhancement, decay lifetimes were measured at various temperatures. Figure 2.7(b) shows the temporal response of the A-exciton (3.485 eV) transition at 10 K both prior to and after microdisk fabrication. Both decays in Fig. 2.7(b) are almost single-exponential. The A-exciton decay lifetime in the microdisks ($\tau = $ 593 ps) is much longer than that for the MQW ($\tau = 53$ ps) and is more representative of a true radiative lifetime. The significantly longer lifetime observed for the A-exciton in the microdisk sample is consistent with the enhancement of the A-exciton transition intensity seen in Fig. 2.7(a).

Optical resonance modes are observed in individually pumped microdisks under high pump intensity. The description of optical resonance modes in a thin dielectric disk involves the satisfaction of Maxwell's equations across a boundary of cylindrical symmetry [31–33]. The fields within the disk are described by Bessel functions, while the evanescent wave outside of the disk is described by Hankel functions. It has been pointed out that the microdisk cavity may support two distinctly different resonant mode types [33]. One mode type is described by Bessel functions $J_m(\chi)$ with $m = -1, 0, 1$ within the cavity. These modes are dominated by photon wave motion along the radial direction of the disks. The equivalent optical path is formed between the edge and the center of the disk, giving an effective round-trip cavity length of $2R$ where R is the radius of the microdisk. This mode consists of radial oscillations of field intensity much like the wavelets formed by a pebble dropped in still water. Another type, known as the whispering gallery (WG) mode [34], is described by Bessel functions $J_m(\chi)$ for large m. The WG mode may be thought of as in-plane propagation around the inside perimeter of the disk which is facilitated by total internal reflection. The effective optical path of $2\pi Rn$ is given by the periodic boundary condition imposed on the circulating wave, with n being the index of refraction of the microdisk.

Strong optical mode behavior was observed in the emission spectra of individually pumped InGaN/GaN MQW microdisks as shown in Fig. 2.8(a). The InGaN/GaN MQWs used for microdisk fabrication consisted of a 30 nm GaN buffer layer followed by a 20-period 45 Å/45 Å In$_x$Ga$_{1-x}$N/GaN ($x \sim 0.15$) MQW. All layers were grown nominally undoped. These mode peaks may be compared with the emission spectrum shown in Fig. 2.8(b) from the InGaN/GaN MQWs without microdisks obtained under equivalent conditions. The three small peaks (3.304 eV, 3.312 eV, and 3.320 eV) observed on the high-energy side of Fig. 2.8(a) exhibit a spacing of 8 meV and are attributed to WG modes. The labeled peaks on the low-energy side of the spectrum (3.174 eV, 3.181 eV, 3.188 eV, and 3.195 eV) are separated by 7 meV and also attributed to the WG mode. The three large peaks (3.219 eV,

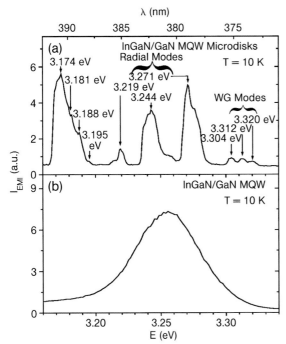

Fig. 2.8 Emission spectra of (a) an individually pumped InGaN/GaN MQW microdisk (of diameter 9.4 μm) and (b) the InGaN/GaN MQW without microdisks. Optical modes of the whispering gallery (WG) and radial types are observed in (a). The observed mode spacings are consistent with the results of calculations. After Ref. [14].

3.244 eV, and 3.271 eV), on the other hand, are spaced by approximately 26 meV and are due to the radial ($m = 0$) modes. The spacing and assignment of the two mode types are discussed in more detail in the following.

Assuming that the disk walls are perfectly conducting so that the field outside of the disk vanishes, in this way, approximate mode positions and spacing are readily derivable. For the radial mode type ($m = 0$), the fields within the disk are described by zeroth-order Bessel functions and the boundary condition for a TM (TE) mode requires that $J_0(kR) = 0$ [or $J_0'(kR) = 0$]. Here, k is the photon wavenumber in the disk and $k = 2\pi/(\lambda/n)$ with λ being the wavelength of the mode. Differentiation is performed with respect to the radial variable. For the case of $m = 0$, the optical modes can be found by noting that in the limit of $kR \gg 1$ (as in this case),

$$J_0(kR) \approx (2/\pi kR)^{1/2} \cos(kR - \pi/4) \tag{2.1}$$

We can thus obtain the eigenmodes for the case $m = 0$ as $J_0(kR) = 0$ [or $J_0'(kR) = 0$], or equivalently for TM modes: $2nR = (p + 3/4)\lambda$ $p = 1, 2, 3, \ldots$ (2.2)

for TE modes: $2nR = (p + 1/4)\lambda$ $p = 1, 2, 3, \ldots$ (2.3)

From Eqs. (2.2) and (2.3), we find that both the TE and TM radial modes exhibit a mode spacing of

$$\Delta\lambda_{\mathrm{rad}}^{\mathrm{TE}} = \Delta\lambda_{\mathrm{rad}}^{\mathrm{TM}} = \lambda^2/2Rn \qquad (2.4)$$

The second type of microdisk cavity mode is the WG mode. This mode has a low loss due to the total internal reflection and thus low threshold for lasing. The effective optical path length of $2\pi Rn$ imposed by the periodic boundary condition results in a WG eigenmode condition of

$$2\pi Rn = m\lambda \quad (m \gg 1) \qquad (2.5)$$

and the mode spacing is given by

$$\Delta\lambda_{\mathrm{WG}} = \lambda^2/2\pi Rn = \Delta\lambda_{\mathrm{rad}}/\pi \qquad (2.6)$$

It is shown in Eq. (2.6) that the radial mode spacing is expected to be larger than the WG mode spacing by a factor of π. For the InGaN/GaN MQW microdisk emission spectrum shown in Fig. 2.8 (a), mode spacings of 8 meV and 26 meV are observed. The expected mode spacing calculated with Eqs. (2.4) and (2.6) and representative disk radius of $R = 4.65\,\mu\mathrm{m}$ and $n = 2.6$ reveals that the observed spacings correspond well to the WG and radial mode types, respectively. From the observed mode spacings of 8 meV and 26 meV for the WG and radial modes, we indeed obtain the ratio of $\Delta\lambda_{\mathrm{rad}}/\Delta\lambda_{\mathrm{WG}} = 3.25 \approx \pi$ as expected from Eq. (2.6).

Stimulated emission and lasing in whispering gallery modes by pulsed optical pumping was also observed at room temperature for microdisk cavities of diameters 50 μm fabricated by reactive ion etching from MOCVD-grown GaN epilayer [35]. The emission spectra from the microdisk show both the WG modes and spectral narrowing. Below the lasing thresholds, stimulated emission with super-linear pump-intensity dependence is observed. A spontaneous-to-stimulated emission transition occurs at a pump intensity that is 10 times lower than that for a GaN sample without a cavity structure. Above the lasing threshold, the pump-intensity dependence is almost linear and gain pinning is also observed. The observed linewidth of the WG modes of individual peaks is as narrow as 0.1 nm. Optically pumped GaN microdisks exhibit three distinctive regimes in the output versus pump intensity characteristics: from (i) sublinear to (ii) superlinear, making the onset of stimulated emission, and then back to (iii) sublinear dependence indicating lasing and gain pinning.

Figure 2.9 shows the emission spectra from a 50 μm diameter GaN microdisk pumped at $\lambda_{\mathrm{p}} = 266\,\mathrm{nm}$ for two different pump intensities I_{p}. At low I_{p}, no mode structure is found. However, at high excitation intensity, individual WG mode peaks are resolved in the stimulated emission spectrum. The large, regularly

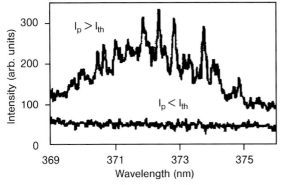

Fig. 2.9 Emission spectra from a 50 μm diameter GaN microdisk pumped at λ_p = 266 nm. Under high excitation, regularly spaced (~0.5 nm) peaks are observed (whispering gallery modes). At low excitation, no peaks are observed. After Ref. [35].

spaced peaks with a free spectral range of $\Delta\lambda \sim 0.5$ nm are WG modes of consecutive angular mode numbers (m and $m + 1$) with the same radial order. The smaller, irregularly distributed peaks are WG modes of different radial orders. From the observed angular mode spacing $\Delta\lambda$, the effective cavity radius R_{eff} is estimated to be around 10.4 μm or $0.42R_0$, where R_0 is the radius of the microdisk.

With the observation of the microdisk cavity effects in GaN, a comparison between GaN microdisks and other conventional III–V semiconductors, say GaAs, can be made in order to provide a guideline for future GaN microdisk lasers. The obvious and attractive distinction of the GaN-based microdisk is the working wavelength range in the blue and ultraviolet. There are also other differences between the conventional and III–nitride microdisk systems. The GaAs-based system benefits from a larger index of refraction, which aids in optical confinement. However, the refractive index difference between GaN and the substrate material (sapphire, $n = 1.8$) is much larger than the index differences found between GaAs-based microdisks and their substrate materials (typically GaAs or InP). This lack of sufficient index difference requires that conventional GaAs microdisks be specially etched to be isolated from the substrate, resulting in a less mechanically stable structure [36]. Perhaps such procedures will not be necessary for the III–nitride/sapphire system. Finally, we note that the cavity quality (Q) for the WG mode is expected to increase with the mode number (m) [36, 37]. By increasing the microdisk radius, one can increase m and thus Q, but the mode density within the spontaneous emission spectrum will also be increased. For a fixed disk radius, the GaN system will have a much greater mode number (m) than the GaAs system because the GaN emission spectrum occurs at a much shorter wavelength. However, for the same disk radius the mode separation (in units of energy) will be comparable. Therefore, with higher achievable mode number for a given mode spacing, the GaN system may prove to benefit more greatly from the microdisk laser geometry than other conventional III–V semiconductors.

2.3.2
Microrings

Microrings of varying sizes were fabricated from the same set of $In_xGa_{1-x}N$/GaN ($x \sim 0.15$) MQWs used for microdisk fabrication [17]. For a microring cavity, one expects to observe only the whispering gallery (WG) mode, which is described by Bessel functions $J_m(\chi)$ for large m. An InGaN/GaN MQW microring array was fabricated and Fig. 2.10(a) shows the schematic diagram. Figure 2.10(b) shows the SEM image of a representative InGaN/GaN MQW microring fabricated by photolithography patterning and ICP dry etching [17].

A PL emission spectrum obtained at 10 K from InGaN/GaN MQWs prior to microring fabrication is plotted in the top portion of Fig. 2.10(c), where the emission line at 3.470 eV originates from the GaN barriers. The emission line at 3.288 eV originates from the InGaN wells. The emission lines at 3.198 eV and 3.108 eV are the one and two longitudinal optical (LO) phonon replicas of the 3.288 eV emission line. An emission spectrum measured at 10 K under high pumping intensity for an individually pumped InGaN/GaN MQW microring with an outer diameter of 3.2 μm and a ring width of 0.7 μm is shown in the bottom portion of Fig. 2.10(c), where strong optical mode behavior is observed. Three strong emission lines at 3.116 eV, 3.170 eV, and 3.224 eV exhibiting a mode spacing of 54 meV are attributed to WG modes. The mode spacing of the WG modes is given by Eq. (2.6), $\Delta\lambda_{WG} = \lambda^2/2\pi Rn$, or in the energy spectrum by

$$\Delta E_{WG} = hc/2pRn \qquad (2.7)$$

where h is the Planck constant, R is the radius of the ring, n is the index of refraction of the InGaN microrings, and λ is the wavelength of light propagating inside the ring cavity.

The calculated mode spacing for $R = 1.6$ μm and $n = 2.6$ is $\Delta\lambda_{WG} = 60$ Å (at $\lambda = 3950$ Å) and $\Delta E_{WG} = 50$ meV, which agrees well with the observed mode spacing. The three peaks at 3.392 eV, 3.444 eV, and 3.494 eV in the high emission energy region, exhibiting a mode spacing of 52 meV and 50 meV, are also WG modes.

The results for an individually pumped MQW microring can be compared with the MQW microdisks. For MQW microdisks, both the radial and the WG modes were observed with two distinct mode spacings and the ratio between the radial and the WG mode spacing is π [12–14]. However, the WG mode is the only type of mode expected from a microring cavity. Furthermore, the full width at half-maximum (FWHM) of the modes observed from the microring is much wider than that from the microdisk mainly due to the finite width of the ring. Unique features of WG modes in a microring cavity include that high Q values can be obtained relatively easily even in a very small mode volume and the number of modes contributing to lasing can be reduced [38].

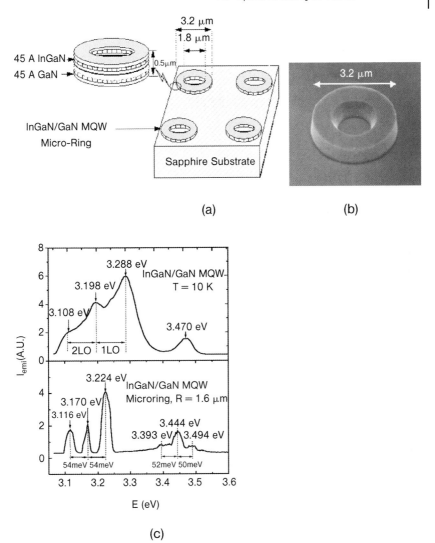

(a) (b)

(c)

Fig. 2.10 (a) Schematic diagram of InGaN/ GaN MQW microrings. (b) SEM image of a representative InGaN/GaM microring fabricated by photolithography patterning and dry etching. (c) PL emission spectrum of an InGaN/GaN MQW sample prior to microstructure formation (top) and optical lasing spectrum of an InGaN/GaN MQW microring (bottom). Compared with the results shown in Fig. 2.8, microring cavities support only WG optical resonance modes. After Ref. [17].

2.3.3
Micropyramids

2.3.3.1 **Optical Properties**
The optical properties of self-organized GaN hexagonal pyramids have been studied [25]. The low-temperature PL spectra of GaN pyramids on GaN/AlN/ silicon and GaN/AlN/sapphire substrates are presented in Figs. 2.11(a) and (b), respectively. For comparison, the PL spectrum of a GaN epilayer on sapphire substrate with AlN buffer layer grown under identical conditions as the GaN pyramids on GaN/AlN/sapphire substrate is also included in Fig. 2.11(c). For the GaN pyramid grown on GaN/AlN/silicon, the main emission line at 3.469 eV is attributed to a neutral donor bound exciton or the I_2 transition. Two other emission lines at 3.422 eV and 3.290 eV are also evident. The emission line at 3.422 eV is very close to the emission line associated with the presence of oxygen impurities in GaN epilayers reported previously [39]. We thus assign the emission line at

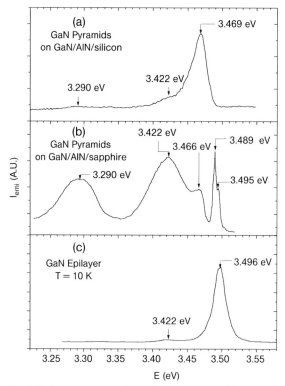

Fig. 2.11 Low-temperature ($T = 10$ K) PL spectra of GaN pyramids overgrown on (a) GaN/AlN/silicon substrate and (b) GaN/AlN/sapphire substrate. (c) The PL spectrum of a GaN epilayer grown under identical conditions. After Ref. [25].

3.422 eV as the recombination between the electrons bound to substitutional oxygen donor impurities and free holes, or the (D^0, h^+) transition.

For the GaN pyramids on GaN/AlN/sapphire substrate [Fig. 2.11(b)], besides the I_2 transition line at 3.466 eV and the impurity-related transition lines at 3.422 eV and 3.290 eV, there also exist transition lines at 3.489 eV and 3.495 eV. In order to identify these two emission lines at 3.489 eV and 3.495 eV, the PL spectra of GaN pyramids on GaN/AlN/sapphire [Fig. 2.11(b)] were compared with that of the GaN epilayer on sapphire grown under identical conditions [Fig. 2.11(c)]. In the GaN epilayer, a strong emission band at 3.496 eV is observed. When the excitation laser intensity, I_{exc}, varied by two orders of magnitude, the spectral peak position (E_p) of the 3.496 eV line in the epilayer on sapphire shifts from 3.496 eV (at $I_{exc} = I_0$) to 3.503 eV (at $I_{exc} = 0.01 I_0$) and the variation of E_p versus I_{exc} follows exactly a 1/3 power law, $E_p \propto I_{exc}^{1/3}$ (not shown). Such a behavior is typical for a band-to-band transition due to the bandgap renormalization effect [40]. The emission lines at 3.489 eV and 3.495 eV in GaN pyramids on GaN/AlN/sapphire were thus assigned to the transitions between the free electrons in the conduction band and free holes in the A and B valence bands, respectively. The 6 meV energy difference is consistent with previous observation and calculation between the A and B valence bands [41].

For GaN pyramids grown on GaN/AlN/sapphire substrate, the emission line at 3.489 eV is about 7 meV below the corresponding peak at 3.496 eV for the GaN epilayer. This 7 meV spectral redshift can be explained by the release of the biaxial compressive strain in the overgrown GaN pyramids. A 1.5 meV blueshift of the band-edge transitions in the laterally overgrown GaN stripes on GaN/AlN/6H-SiC(0001) substrate with respect to that of the underlying GaN epilayer has been previously reported [42]. This is expected since the GaN epilayer on 6H-SiC(0001) substrates is subject to a biaxial tensile strain [43]. However, in our case with a sapphire substrate, it corresponds to a biaxial compressive strain. Its release in the GaN pyramids leads to a redshift of the spectral peak position. The magnitude of the strain in the GaN epilayer can also be estimated. The 7 meV redshift corresponds to a release of the ε_{zz} value of about 0.05% (denoting the magnitude of the uniaxial strain along the c-axis) [44] in the GaN pyramids with respect to that of the GaN epilayer.

Diffusion of Si and O impurities from the SiO_2 mask during pyramid overgrowth should leave an impurity distribution in the pyramids with fewer Si and O impurities close to the top of the pyramids. In order to check the crystalline quality and purity in different parts of the pyramids, two different configurations have been employed to collect PL from the overgrown pyramids on GaN/AlN/sapphire as illustrated in the insets of Fig. 2.12. In both configurations, the incident laser beam was perpendicular to one of the six surfaces of the pyramid. PL was collected along the central axis (or one of the surfaces) of the pyramid as shown in Fig. 2.12(a) [Fig. 2.12(b)] in such a way that the PL from the top (or base) part of the pyramids dominates. Comparing the PL results shown in Fig. 2.12, the intrinsic band-to-band transitions relative to the extrinsic band-to-impurity

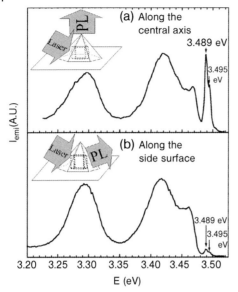

Fig. 2.12 Low-temperature (10 K) PL spectra of a GaN pyramid overgrown on a GaN/AlN/sapphire substrate collected along (a) the central axis and (b) one of the six surfaces of the pyramid. After Ref. [25].

transitions are significantly enhanced in the top of the pyramids [configuration Fig. 2.12(a)]. The absolute emission intensity of the band-to-band transitions in the top part of the pyramid [Fig. 2.12(a)] is also much higher than that in the base part. These results imply that the crystalline quality and purity in the top part of the pyramids is higher than that in the base part. These results are consistent with those reported elsewhere [21]. It was shown there that dislocation diminishes above approximately one-third of the pyramid height within a pyramidal volume.

The micropyramids could potentially be used for the development of two-dimensional laser arrays. Optically pumped laser action was observed in GaN pyramids grown on (111) Si by selective lateral overgrowth of MOCVD [24]. The integrated emission intensity of both spontaneous and lasing peaks was measured as a function of excitation power density. The effects of pyramid geometry and short-pulse excitation on the multimode nature of laser oscillations inside of the pyramids were considered.

Figure 2.13(a) shows the room-temperature emission spectra at different pump densities near the lasing threshold. At excitation densities below the lasing threshold, only a spontaneous emission peak with a FWHM of approximately 14 nm is present, shown as the bottom plot. As the pump density is increased to values slightly above the lasing threshold, several equally spaced narrow peaks with FWHMs of less than 0.3 nm appear. The position of the peaks remains the same

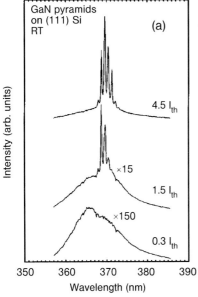

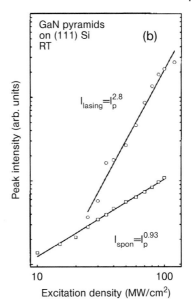

Fig. 2.13 (a) Emission spectra of a GaN pyramid measured at room temperature under different levels of excitation intensity above and below the lasing threshold. At pump densities below the lasing threshold, only a 14 nm wide spontaneous emission peak is present, whereas at excitation levels above the lasing threshold, multimode laser action is observed. The mode spacing corresponds to a microcavity of 58 μm in perimeter. (b) Plot of the peak intensity as a function of excitation densities. The intensity of the spontaneous emission peak (open squares) increases almost linearly with excitation power. The lasing peak intensity (open circles) exhibits a strong superlinear increase as the pump intensity is raised. The solid line represents a linear fit to the experimental data. After Ref. [24].

as the excitation density is further increased. Due to the pump density dependence of the effective gain profile, the maximum intensity mode tends to hop to modes with lower energy as shown in the spectrum plotted at the top of Fig. 2.13(a).

The intensity dependence of the spontaneous and lasing peaks as a function of excitation power was measured. It was observed that the intensity of the spontaneous emission peak increases almost linearly with pump density ($I_{spon} = I_p^{0.93}$), as illustrated in Fig. 2.13(b) (open squares). On the other hand, the intensity of the lasing peaks experiences a strong superlinear increase ($I_{lasing} = I_p^{2.8}$) with excitation power (open circles).

Selective MOCVD growth of InGaN quantum dots (QDs) on the top of hexagonal GaN pyramids has also been explored [26]. Selective growth of three periods of InGaN MQWs at 720 °C was followed by the formation of hexagonal GaN pyramids. The growth times for the InGaN well and $In_{0.02}Ga_{0.98}N$ barrier materials were controlled to give 2.4 nm and 4.1 nm thicknesses, respectively, in planar growth. Then, a 20 nm thick $In_{0.02}Ga_{0.98}N$ layer was deposited. InGaN QD structures were

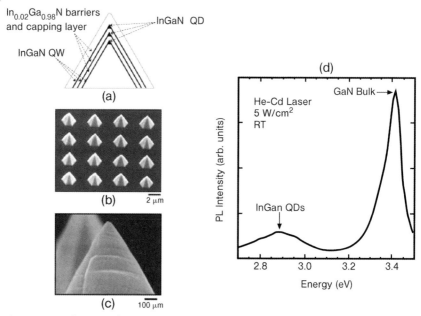

In$_{0.02}$Ga$_{0.98}$N barriers and capping layer

InGaN QD

InGaN QW

(a)

(b) 2 μm

(c) 100 μm

(d)

He-Cd Laser
5 W/cm^2
RT

GaN Bulk

InGan QDs

PL Intensity (arb. units)

Energy (eV)

Fig. 2.14 (a) Schematic of an InGaN QD formed on the top of a hexagonal GaN pyramid. SEM pictures of the sample: (b) bird's-eye view and (c) cross-section. (d) Room-temperature PL spectrum of the sample. After Ref. [26].

then formed at the top of the hexagonal pyramids, as illustrated schematically in Fig. 2.14(a). Figure 2.14(b) shows a SEM bird's-eye view of the final structures. Figure 2.14(c) shows a cross-sectional image after the sample was cleaved. Figure 2.14(d) shows the PL spectrum of the sample. As shown in Fig. 2.14(d), the peak energy of the emission attributed to the InGaN QD structures is 2.88 eV (430 nm), and the FWHM is 290 meV. The spectrum is broad due the presence of many QD structures that were excited. The intensity of the InGaN QD PL peak is smaller than that of the GaN emission peak at 3.42 eV, because the volume of InGaN QDs is much smaller than that of GaN bulk. To identify the regions giving the emission at 430 nm, micro-PL intensity images were recorded at room temperature, using a conventional optical microscope equipped with an objective lens of which the nominal magnification factor was 100 and the numerical aperture (NA) was 1.3 when immersion oil was used, giving a typical spatial resolution as high as 150 nm. Figure 2.15(a) shows an image obtained only through a barrier filter of which the cutoff wavelength was 400 nm, without the excitation filter or bandpass filter, consisting mainly of reflected light. In Fig. 2.15(a), hexagonal shapes can be seen very clearly. Figure 2.15(b) shows a micro-PL intensity image of PL around 430 nm, accumulated over 10 s. The same area was observed in Fig. 2.15(b) as in Fig. 2.15(a). It can be seen that the 430 nm PL emission in Fig. 2.15(b) is only from the top of the pyramids, not from the edges. In the cross-sectional profile of PL

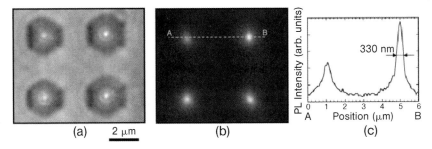

(a)   2 μm   (b)   (c)

Fig. 2.15 (a) Reflective image, (b) micro-PL intensity image
at the emission wavelength of InGaN QDs (430 nm), and
(c) cross-sectional profile of PL intensity along the line AB in
(b). The results show that the emission at 430 nm is only
from the tops of the hexagonal pyramids and hence confirm
the formation of InGaN QDs on the tops of the pyramids.
After Ref. [26].

intensity in Fig. 2.15(c), the FWHM of 330 nm is comparable to the spatial resolution. Such a narrow width of emission areas indicates that the emission originates
from InGaN QD structures embedded in the $In_{0.02}Ga_{0.98}N$ matrix and hence the
existence of InGaN QDs on the top of the hexagonal pyramids.

2.3.3.2 Pyramidal Microcavities
The optical mode behaviors of GaN hexagonal pyramidal microcavities fabricated
by MOCVD selective epitaxial overgrowth were also studied [45]. These microscale
pyramids are highly efficient microcavities due to the extremely smooth facet surfaces formed by self-organization without any processing damage. Optical resonance modes in the GaN pyramidal microcavities can be observed at a pumping
intensity that is several orders of magnitude lower than that in the III–nitride
MQWs microdisk cavities. These results suggest that self-organized microcavities
formed by selective epitaxial overgrowth can be further developed for the realization of GaN microcavity laser arrays with minimal optical losses as well as a simplified processing procedure that eliminates the need for etching the crystals. A
micropyramid is a true 3D cavity, which can support several different types of
optical resonance modes. These novel and unique pyramidal microcavities open
new avenues for optoelectronic device applications.

Figure 2.16 shows low-temperature (10 K) PL emission spectra of a GaN pyramid
pumped under (a) low excitation intensity and (b) high excitation intensity. The
observed optical mode patterns shown in Fig. 2.16(b) are quite complicated, but
can be classified into three different groups with different mode spacings. The
first group includes the emission peaks at 3555 Å, 3565 Å, and 3576 Å, giving a
mode spacing of $\Delta\lambda_1^{obs} = 10$ Å. The second group includes the emission peaks at
3698.5 Å, 3703.5 Å, and 3708.5 Å as well as at 3681 Å and 3686 Å, giving a mode
spacing of $\Delta\lambda_2^{obs} = 5$ Å. The third group includes emission peaks at 3590 Å, 3596 Å,

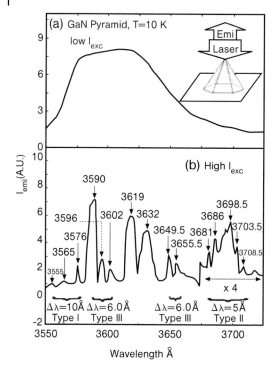

Fig. 2.16 (a) PL emission spectrum of GaN pyramids under a low pump intensity. (b) Optical resonance mode behavior of an individually pumped pyramid under a high pump intensity. After Ref. [45].

and 3602 Å as well as those at 3649.5 Å and 3655.5 Å, giving a mode spacing of $\Delta\lambda_3^{obs} = 6.0$ Å. Notice that $\Delta\lambda_1^{obs}/\Delta\lambda_2^{obs} = 2$ and $\Delta\lambda_1^{obs}/\Delta\lambda_3^{obs} = 1.7$.

Figure 2.17(a) shows a schematic diagram of a GaN pyramid microcavity and provides the dimensions of the microscale pyramids ($a = 8.0\,\mu m$ and $h = 1.63a$). The angle between the facets and the SiO$_2$ basal plane is $\alpha = 62°$. To study the optical resonance mode behaviors, a UV-transmitting objective was used in a confocal geometry to optically pump a single pyramid normal to the sample surface and to collect the light emission in the direction of the surface normal. Focused beam spot diameters as small as 2 μm could be achieved with the UV objective lens [45].

The observed mode spacing can be understood from the following calculation, which also provides a clear picture for the formation of the optical cavities inside a micropyramid. The first type can be identified quite easily from Fig. 2.17(a). Two opposite facets together with the SiO$_2$ basal plane form one cavity, i.e., between planes AEB, ADC, and EBCD [see Fig. 2.17(a)]. The side view of such a cavity or its projection on a plane normal to these two facets (AEB and ADC) is presented in the top half of Fig. 2.17(b). The six facets of the pyramids together with the SiO$_2$

Cavity Type I

(a)

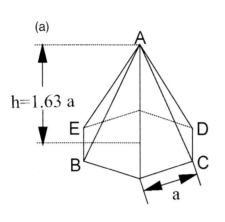

(b)

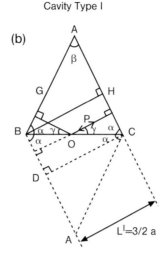

Fig. 2.17 (a) Schematic of a GaN pyramid. (b) Optical path of a Fabry–Pérot-like cavity formed in a GaN hexagonal pyramidal microcavity determined by an optical ray-tracing method. The effective optical path of this first type of optical resonance mode is about 3 *an*. After Ref. [45].

basal plane [see Fig. 2.17(a)] form three pairs of cavities of this type. A light beam (line P) propagating in a direction normal to one of the facets (facet AC) will be reflected by the basal plane BC. The reflected light beam is also normal to the opposite facet (facet AB) since $\angle$ABC = $\angle$ACB = α. A cavity is thus formed with a cavity length of about $L_c^1 = (3/2)a$, where a is the side length of the pyramids. Three facets in a pyramidal microcavity form this cavity, which is similar to the Fabry–Pérot cavity formed by two parallel facets. This cavity gives a standing wave of light inside a pyramid with a total optical path of about $2nL_c^1 (= 3an)$ with n being the index of refraction of GaN.

A more intriguing way to visualize this cavity in the pyramids can be achieved by the following method. Instead of reflecting the light beam (line P) at point O on the basal plane, one allows the light beam to propagate continuously on a straight line and the pyramid to reflect about the basal plane BC [see the lower half of Fig. 2.17(b)]. By doing so, the total active area involved in this particular type of optical resonance mode can be easily determined, which is the lower half area of the facets, between planes HC and GB (the thicker lines). This is a relatively large area and thus gives a strong signature of this type of optical mode in the emission spectrum as shown in Fig. 2.16(b). The cavity type has the shortest cavity length and hence the largest mode spacing, $\Delta\lambda_1 = 10\,\text{Å}$. This method gives a very handy way to find out other types of optical resonance modes in the pyramids.

By employing the same method as described above, the second type of cavity can be readily identified and is schematically illustrated in Fig. 2.18(a). It is easy

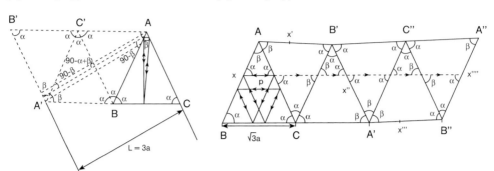

Fig. 2.18 (a) Optical path of a Fabry–Pérot-like cavity formed in a GaN hexagonal pyramidal microcavity determined by an optical ray-tracing method. The effective optical path of this second type of optical resonance mode is about $6\,an$. (b) A WG-like cavity formed in a GaN pyramidal microcavity with a total optical path of about $3(3)^{1/2}\,an$, where n is the refractive index of GaN. After Ref. [45].

to prove [following all the angles in Fig. 2.18(a)] that the light beam, which is normal to the plane AC, is also normal to the plane A′B′. It thus forms a cavity with cavity length of about $L_c^{II} = 3a$, twice the cavity length of the first optical mode type. The mode spacing of the second type of resonance mode is thus one-half of that of the first type, which is observed experimentally in Fig. 2.16(b), $\Delta\lambda_2^{obs} = 5\,\text{Å}$. Again, the thicker lines on the top portion of the pyramid represent the size of this type of cavity. The signal of this optical mode type is weaker because more reflection surfaces were involved, which is illustrated in Fig. 2.16(b). Resonance modes in Fig. 2.18(a) are also a standing wave.

One can also visualize the existence of the third type of optical mode inside the pyramids. The dotted line between x and x″″ in Fig. 2.18(b) represents a straight light beam reflecting six consecutive times from ABC to A″B″C″. It is clear that the point x″″ on the plane A″B″ is the same point as x on the plane AB. With light beam P parallel to the plane BC, it is easy to prove that the plane AB is parallel to the plane A″B″ by following all the angles indicated along the light beam xx″″ by using the relation $2\alpha + \beta = 180°$. The optical path thus takes a complete cycle, which has a great similarity to the WG modes in the microdisk and microring cavities. However, this traveling wave propagates in the direction perpendicular to the growth plane, while the typical WG modes propagate inside the growth plane. One of the actual optical paths is indicated in Fig. 2.18(b) by the solid line inside the pyramid ABC. The total optical path for this type of mode is about $3(3)^{1/2}an$.

The optical loss of this mode is very low since the light beam is reflected by total internal reflection. A total of six internal reflections are involved for the optical path to complete a cycle.

The resonant modes of wavelength λ in the pyramidal microcavities can be calculated by

$$Ln' = m\lambda \quad (m \text{ is an integer}) \tag{2.8}$$

with mode spacing

$$\Delta\lambda = \lambda^2/Ln' \tag{2.9}$$

where Ln' is the total optical path of different resonance modes and n' is the GaN effective index of refraction under a given excitation condition, which was taken as $n' = 2.65$. From Figs. 2.17 and 2.18, the total optical paths of the first, second, and third optical mode types are about $3an'$, $6an'$, and $3(3)^{1/2}an'$, with mode spacings of $\Delta\lambda_1^{cal} = 20.4\,\text{Å} \approx 2\Delta\lambda_1^{obs}$, $\Delta\lambda_2^{cal} = 10.2\,\text{Å} \approx 2\Delta\lambda_2^{obs}$, and $\Delta\lambda_3^{cal} = 11.8\,\text{Å} \approx 2\Delta\lambda_3^{obs}$, respectively. The ratios are expected to be $\Delta\lambda_1^{cal}/\Delta\lambda_2^{cal} = 2$ and $\Delta\lambda_1^{cal}/\Delta\lambda_3^{cal} = 1.7$. We see that the calculated mode spacing ratios agree very well with the observation of $\Delta\lambda_1^{obs} = 10\,\text{Å}$, $\Delta\lambda_2^{obs} = 5\,\text{Å}$, and $\Delta\lambda_3^{obs} = 6.0\,\text{Å}$ and the ratios of $\Delta\lambda_1^{obs}/\Delta\lambda_2^{obs} = 2$ and $\Delta\lambda_1^{obs}/\Delta\lambda_3^{obs} = 1.7$. The resonance modes can also be determined precisely by solving Maxwell's equation by satisfying the boundary conditions imposed by the symmetry properties of pyramids. The allowed eigenmodes resulting from such a calculation should be exactly double those obtained from the ray-tracing method developed here.

Arrays of InGaN QW microcone cavities were also fabricated by ion beam etching [46]. Optical resonance modes from a single microcone could be clearly observed in the PL spectra at temperatures up to 200 K. The optical ray-tracing method described above has been used to calculate the four main types of optical resonance cavities in the microcone, including two Fabry–Pérot modes as well as two WG modes. The calculated mode spacing agrees well with the observed resonant mode spacing. The advantages of microcone cavities compared with other microcavities were also discussed.

2.3.4
Submicron Quantum-Well Waveguides

It was found that, when the width of the AlGaN/GaN QW waveguides such as those shown in Fig. 2.6 was reduced to below 0.7 μm, the PL emission peak position and linewidth of the localized exciton emission from the QW waveguides varied systematically with orientation of the waveguides and followed the six-fold symmetry of the wurtzite structure [27]. This is illustrated in Fig. 2.19. Figure 2.19(a) only shows the PL spectra from an unetched portion of the sample (top) and the PL spectra from waveguides oriented at 20° and 60° with respect to the *a*-axis of GaN, as schematically illustrated in Fig. 2.19(b). The emission peaks in

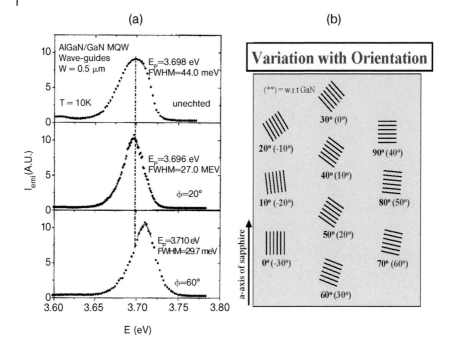

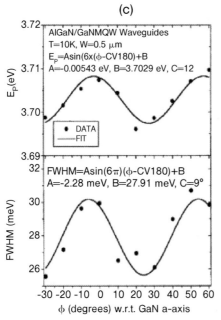

Fig. 2.19 (a) Low-temperature (10 K) CW PL spectra from AlGaN/GaN MQW waveguides of different orientations. For clarity, the PL spectrum from an unetched portion of the sample (top) and the PL spectra from waveguides oriented at $\Phi = 20°$ (center) and $\Phi = 60°$ (bottom) with respect to the a-axis of GaN are shown. (b) Schematic of the AlGaN/GaN QW waveguides with different orientations. (c) Variation of the spectral peak positions (E_p) and full width at half-maximum (FWHM) of the PL emission line at 10 K with respect to the a-axis of GaN. The solid line is a sinusoidal fit of the data with six-fold symmetry of hexagonal structure. After Ref. [27].

these spectra are attributed to localized exciton recombination in the well regions of the waveguide structures. It was found that the linewidth of the spectrum from the unetched region is much broader than that of the spectrum from the waveguides. The peak position E_p and FWHM depend on the orientation, as can be seen in Fig. 2.19(a). Both E_p and FWHM of the emission line versus the waveguide orientation Φ relative to the a-axis of GaN are shown in Fig. 2.19(c). As shown, there is a definite periodicity of 60° in E_p and FWHM, both varying sinusoidally as $A \sin[6\pi(\Phi - C)/180] + B$, where A, B and C are variables. It can be seen that E_p and FWHM are both maximum for orientations roughly parallel to 0° and 60° and both minimum for orientations roughly parallel to −30° and 30°. Between the two extremes, E_p changes by about 11 meV while FWHM changes by about 4.6 meV.

The observed anisotropic optical properties with the six-fold symmetry in the nitride quantum-well plane can be understood by the anisotropic diffusion of photoexcited carriers and excitons for waveguides along different orientations. For fixed excitation laser intensity, due to the band filling effect, carrier or exciton density will be higher for the case of either lower diffusion coefficient or less potential fluctuation. These differences could be the source of the anisotropy of the exciton/carrier diffusion coefficient in the quasi-1D waveguide structures. At 0°/60° orientations, the slow carrier or exciton diffusion leads to a maximum for both E_p and FWHM. Faster diffusion occurs along the −30°/30° resulting in E_p and FWHM both being minimum.

The most intriguing result obtained from these waveguides is that there is a difference in optical properties of submicron structures, shown by the periodic variation in the peak energy E_p and FWHM of the spectra from MQW waveguides at different crystal orientations. This difference is more pronounced in smaller structures. The major implication of this result is that, in photonic and electronic devices where structures of submicron or nanoscale sizes are involved, there will be differences in exciton or carrier dynamics. The differences arising from the choice of orientation will result in significant effects in the associated devices. Such devices include field effect transistors (FETs), optical waveguides, photodetectors and ridge-guide laser diodes. Therefore in the design of these devices, proper attention must be paid to the choice of the orientation of the associated submicron structures.

The light propagation properties in the AlGaN/GaN MQW waveguides were also studied [28, 29]. The AlGaN/GAN MQW waveguide sample was placed normal to the incident light and the emitted photons were collected from a direction parallel to the sample surface, as schematically shown in Fig. 2.20. The incident laser beam was focused to a spot size of about 2 μm on a single waveguide using a UV-transmitting objective lens of focal length 3 mm. The distance d (μm), from the edge of the waveguide closest to the collecting slit of the monochromator, defines the position of the incident excitation laser spot focused on the waveguide.

It was found that the emission intensity of the 3.585 eV line, attributed to the localized exciton recombination in the quantum-well regions of the waveguide structure [27, 29, 47, 48], decreased with increasing d. This result is shown in

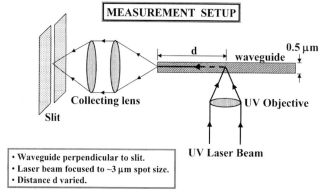

Fig. 2.20 Schematic of the time-resolved PL measurement setup for light propagation studies in waveguides. The value *d*, measured from the edge of the waveguide closest to the slit, defines the position of the focused laser spot on the waveguide. After Ref. [28].

(a)

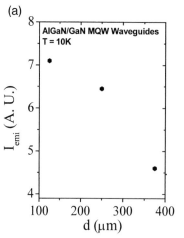
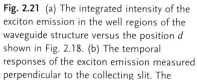

(b)

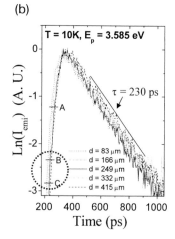

Fig. 2.21 (a) The integrated intensity of the exciton emission in the well regions of the waveguide structure versus the position *d* shown in Fig. 2.18. (b) The temporal responses of the exciton emission measured perpendicular to the collecting slit. The responses were collected from five different positions *d* of the same waveguide with fixed excitation conditions. From the time delays, an average propagation speed of $(1.26 \pm 0.16) \times 10^7 \, \mathrm{m \, s^{-1}}$ of light in the waveguide was determined. After Ref. [28].

Fig. 2.21(a). The decrease in exciton emission intensity with increase in *d* can be understood in terms of light propagation loss in the waveguides that occurs in many different forms including scattering at the walls of the waveguides and reabsorption. At the spot on the waveguide where initial excitation occurs, carriers and excitons are generated. They recombine and emit light that is laterally confined to propagate along the waveguide. PL emitted at the excitation spot will traverse the distance d along the waveguide before being detected. The greater the distance the photons have to travel, the more the chance of loss through scattering or reabsorption leading to a decrease in the integrated emission intensity, as seen in Fig. 2.21(a).

Figure 2.21(b) shows the temporal response of the exciton PL emission at 3.585 eV, corresponding to the spectral peak position for the waveguide aligned perpendicular to the collecting slit. The temporal PL responses were collected from five different positions d on the same waveguide with fixed excitation conditions. The responses for different d values are similar in slope in both the rise part and the decay part. The decay can be fitted quite well with a single exponential giving a lifetime of 230 ± 2 ps. However, there is a systematic increase in time delay in the initial PL signal buildup as d is increased. From this time delay, an average velocity of $(1.26 \pm 0.16) \times 10^7 \, \mathrm{m \, s^{-1}}$ was obtained, which is the propagation speed of generated photons, with energy corresponding to the well transitions in the MQW. Using an approximate value of refractive index $n = 2.67$ for GaN [49], the speed of light in the waveguide is estimated to be $c/n = 1.12 \times 10^8 \, \mathrm{m \, s^{-1}}$. This is an order of magnitude greater than the velocity of the generated photons in the waveguides that we have determined.

In direct bandgap semiconductors including the III–nitrides, polaritons, the coupled mode of photons and excitons, is the normal mode of propagation of light in semiconductors in the neighborhood of exciton resonant energy [50]. In a quantum-well system, the coupling between excitons and photons to form excitonic polaritons is further enhanced due to the quantum confinement effect [51, 52]. A quasi-one-dimensional structure such as the waveguide structure we have studied is expected to show even more stable excitonic polaritons because of the increased oscillator strength of excitons [53, 54]. It is thus expected that the light generated in the waveguides propagates in excitonic polariton mode. In this mode, the propagation velocity has strong energy dependence particularly in the knee region of the polariton dispersion curve, and is typically much smaller than the speed of light in semiconductors. The reduced propagation speed of the polaritons in the waveguide is expected since the coupling between the excitons and photons occurs at an energy corresponding to exciton transitions in the MQW. The average speed obtained is a measure of the propagation speed of polaritons inside the waveguide. Polaritons with speeds three or four orders of magnitude slower than the speed of light in different semiconductors in the bottleneck region has previously been observed [50, 55, 56].

These results shed light on the understanding of the fundamental properties behind the propagation of light in III–nitride semiconductors. This information is important for many device applications. For example, when the ridge-guide laser diode is used as a read/write laser source in digital versatile disks (DVDs), the ridge width has to be reduced to micron dimensions in order to obtain the fundamental transverse modes necessary to collimate the laser light to a small spot [57, 58]. The knowledge of the speed at which light is propagated along such a device is basic to its design for improved operation.

2.3.5
Optically Pumped Vertical-Cavity Surface-Emitting Laser Structures

Vertical-cavity surface-emitting laser (VCSEL) structures based on III–nitrides have also attracted increasing attention because they are expected to surpass conventional blue nitride LDs in many applications. In particular, the use of two-dimensional arrays of blue VCSELs would drastically reduce the read-out time in high-density optical storage (DVD) and increase the scan speed in high-resolution laser printing technology. Optically pumped laser action has been demonstrated at blue wavelength at room temperature in GaN VCSELs [59]. The microcavity was formed by sandwiching InGaN MQWs between nitride-based and oxide-based quarter-wave distributed Bragg reflectors (DBRs). Using a microcavity with a high Q-factor of 500, lasing action was observed at a wavelength of 399 nm and confirmed by a narrowing of the linewidth in the emission spectra from 0.8 nm below threshold to less than 0.1 nm (resolution limit) above threshold. These results suggest that practical blue VCSELs can be realized in nitride-based materials systems.

Figure 2.22 shows (a) an SEM image of a two-dimensional array of GaN-based VCSEL structure and (b) a cross-sectional image of the VCSEL structure observed by transmission electron microscopy. Disk-shaped VCSEL structures 18 μm in diameter are formed by reactive ion etching and arrayed in a two-dimensional matrix with 22 μm spacing. As can be seen in the cross-sectional view of the VCSEL structure in Fig. 2.22(b), a 2.5λ cavity containing $In_{0.1}Ga_{0.9}N$ MQWs as the active region is sandwiched between two highly reflective mirrors: the nitride-based DBR and an oxide-based one. The nitride DBR consists of 43 pairs of 38 nm GaN (dark layers) and 40 nm $Al_{0.34}Ga_{0.66}N$ (bright layers). The oxide DBR consists of 15 pairs of 48 nm ZrO_2 (dark layers) and 68 nm SiO_2 (bright layers). The peak reflectivity was 98% and 99.5% for the nitride and oxide DBRs, respectively.

A typical microscopic emission image from a single VCSEL structure below threshold is shown in Fig. 2.23(b). As shown in Fig. 2.23(c), the spontaneous emission (below threshold) of the VCSEL structure was modified by the microcavity effects. The main peak at $\lambda = 399$ nm is a resonance cavity mode, and the two other peaks at $\lambda = 389$ and 408 nm are due to modulation produced by the transmission spectrum of the nitride DBR. The spectral linewidth of the cavity mode may be as narrow as 0.8 nm as shown in Fig. 2.21(c), giving a Q-factor of 500, which corresponds to a cavity with 98% reflecting mirrors, neglecting internal loss. It is estimated that the optical gain of the InGaN QW at threshold is larger than 890 cm^{-1}, which is comparable with that of InGaAs/GaAs QWs.

The dependence of the laser output power on the input excitation power measured at room temperature, as shown in Fig. 2.24(a), reveals clearly a threshold at a pump energy of $E_{th} = 43$ nJ, which corresponds to 10 mJ cm^{-2}. The electron sheet density at threshold is estimated to be around 2×10^{12} to 4×10^{12} cm^{-2}. High-resolution measurements of the emission spectra, as shown in Fig. 2.24(b), from the VCSEL for increasing excitation powers show a transition from the 0.8 nm

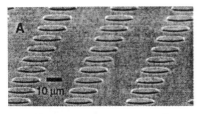

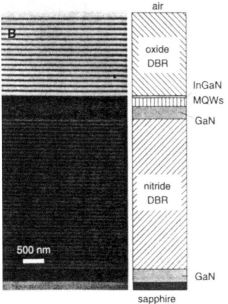

Fig. 2.22 Optically pumped lasing action has been demonstrated at blue wavelength in a vertical-cavity surface-emitting laser structure based on InGaN/GaN MQWs. (a) SEM image of a 2D array of GaN-based VCSEL structures. (b) TEM cross-section image of the VCSEL structure. After Ref. [59].

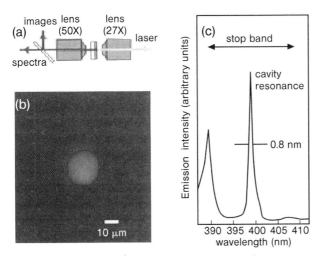

Fig. 2.23 (a) Experimental setup to measure microscopic emission images. (b) Microscopic emission image from a single VCSEL structure below threshold measured at room temperature, with a He–Cd laser ($\lambda = 325$ nm) as the excitation source. (c) Spectrally resolved spontaneous emission of (b). The arrow indicates the stopband of the nitride DBR. After Ref. [59].

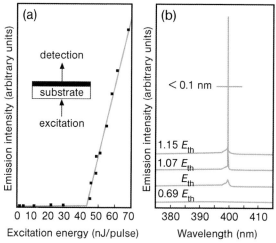

Fig. 2.24 Emission from a single VCSEL structure measured at room temperature with an excitation source of a dye laser ($\lambda = 367$ nm), pumped by a nitrogen laser. (a) Emission intensities accumulated for 180 pulses are plotted as a function of excitation energy per pulse. (b) Emission spectra from the VCSEL structure at various excitation powers ($E_{th} = 43$ nJ threshold energy). After Ref. [59].

width spontaneous emission peaks spectrally filtered by microcavity below threshold to very sharp emission peaks above threshold. Such a spectral narrowing is direct evidence of lasing action.

2.4
Current-Injection Microcavity Light-Emitting Diodes

2.4.1
Microdisk Cavity LEDs

Recently, a III–nitride microscale LED (μ-LED) technology has been developed [7, 60, 61]. In this section, we summarize the fabrication and characterization of InGaN/GaN QW μ-LEDs, including device fabrication processes, I–V characteristics, light output power, electroluminescence (EL) spectra, and comparisons between conventional broad-area LEDs and μ-LEDs [60, 61].

The μ-LEDs were fabricated from InGaN/GaN QW LED wafers. A schematic diagram of μ-LEDs fabricated by the authors' laboratory is shown in Fig. 2.25(a). A SEM image of an array of fabricated μ-LEDs is shown in Fig. 2.25(b) for one representative size. The QW LED wafers comprised 3.5 μm of Si-doped GaN, 0.1 μm of Si-doped superlattice consisting of alternating layers of 50 Å/50 Å AlGaN/GaN, 50 Å of Si-doped GaN, a 20 Å undoped InGaN active layer, 0.14 μm of Mg-

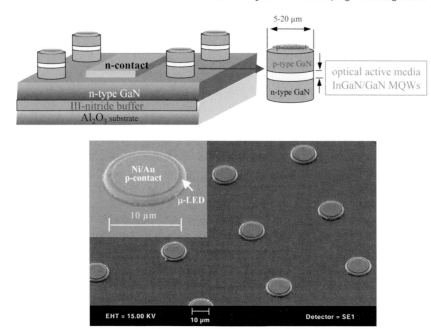

Fig. 2.25 (a) Schematic diagram and (b) SEM image of
InGaN/GaN quantum-well microscale LEDs (μ-LEDs). After
Ref. [60].

doped superlattice consisting of alternating layers of 50 Å/50 Å AlGaN/GaN, and
0.5 μm of Mg-doped GaN epilayer. The LED structures were treated by a rapid
thermal annealing at 950 °C for 5 s in nitrogen. This process produced p-layer
concentrations of 5×10^{17} cm^{-3} (hole mobility 12 cm^2 V^{-1} s^{-1}) and n-layer
concentrations of 1.6×10^{18} cm^{-3} (electron mobility 310 cm^2 V^{-1} s^{-1}). By incorporating the AlGaN/GaN superlattice structure into LED device layers, the p-type con-
centration was enhanced from 2×10^{17} cm^{-3} to 5×10^{17} cm^{-3}. Microscale LEDs with
size varying from 5 to 20 μm were fabricated by photolithographic patterning and
inductively coupled plasma (ICP) dry etching. ICP etching produced very smooth
top and edge surfaces. Bilayers of Ni(20 nm)/Au(200 nm) and Al(300 nm)/Ti(20 nm)
were deposited by electron-beam evaporation as p- and n-type ohmic contacts. The
p- and n-type contacts were thermally annealed in nitrogen ambient for 5 min at
650 °C, respectively. The inner (outer) circle with a diameter of about 8 μm (12 μm)
shown in the inset of Fig. 2.25(b) is the image of p-type contacts (μ-LEDs). The
emission wavelengths of μ-LEDs can be varied from purple to green (390 to
450 nm) by varying the indium (In) content in the InGaN active layers.

In order to characterize individual μ-LEDs, a dielectric layer was deposited by
e-beam evaporation after the formation of μ-LEDs for the purpose of isolating p-
type contacts from the etch-exposed n-type layer. Figure 2.26(a) shows atomic force
microscope (AFM) images of a fabricated μ-LED. As can be seen from Fig. 2.26(a),
the p-type contact was deposited onto the top p-layer by opening a hole through

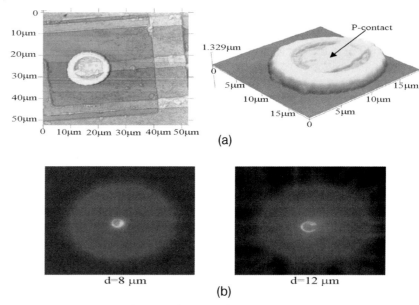

Fig. 2.26 (a) AFM images of a III–nitride μ-LED of diameter
$d = 12\,\mu m$. (b) Optical microscope images of two μ-LEDs
($d = 8$ and $12\,\mu m$) in action. After Ref. [61].

the insulating dielectric layer. The size of the p-type contact is about $4\,\mu m$ in
diameter in this case. Figure 2.26(b) shows optical microscope images, taken from
the top (p-type contact side), of two representative InGaN/GaN QW μ-LEDs with
diameters $d = 8$ and $12\,\mu m$ in action under an injected current of $2\,mA$. The p-type
contacts on the top layers are also visible in Fig. 2.26(b).

The I–V characteristics of microdisk LEDs of varying sizes and a conventional
broad-area LED ($300 \times 300\,\mu m^2$) fabricated from the same wafer are plotted in Fig.
2.27(a). It is clearly seen that the turn-on voltages for individual μ-LEDs are larger
than that of the broad-area LED. Among the different sizes of μ-LEDs, the turn-on
voltage increases with decreasing μ-LED size. The slope of the log I versus V plot
in Fig. 2.27(a) reflects the ideality factor, n (= 1/slope). It is clear that the ideality
factor of μ-LEDs ($n = 18.5$) is larger than that of the broad-area LED ($n = 6.4$). There
is only a weak size dependence of ideality factor for the microdisk LEDs. The larger
ideality factor reflects the enhanced nonradiative recombination in μ-LEDs, which
is most likely a result of enhanced surface recombination around the edge of the
disk of μ-LEDs.

The size dependence of the μ-LED power output has been investigated. Figure
2.27(b) plots the output power versus input power measured from the sapphire
substrate side for three unpackaged $408\,nm$ μ-LEDs of different sizes. Heating
effects become more prominent as the size of the μ-LEDs decreases. For μ-LEDs
with $d = 12\,\mu m$, the output power increases almost linearly with input power in
the entire measured range. However, for μ-LEDs with $d = 8\,\mu m$, the output power

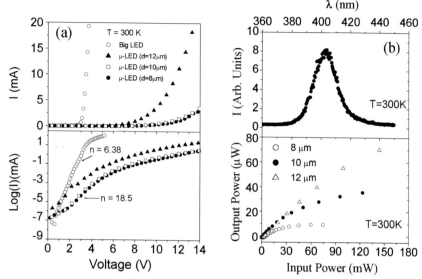

Fig. 2.27 (a) The *I–V* characteristics of μ-LEDs of varying sizes (*d* = 8, 10, and 12 μm) and a broad-area LED (300 × 300 μm²), in linear (top) and semilogarithmic (bottom) plots. (b) The *E–L* emission spectrum of a purple μ-LED (top) and output power versus input power (*L–I*) plot of μ-LEDs of different sizes (bottom). After Ref. [61].

saturates at about 10 μW for input power above about 45 mW. As expected, heat dissipation is more difficult in μ-LEDs with reduced sizes, which causes power output saturation.

These μ-LEDs have many potential applications, including short-distance optical communications, chip-scale optical sources for biological and chemical agent detection, etc. For these applications, a pulsed light source may be preferred and hence the operating speed is one of the important parameters. Time-resolved EL was employed to study the response time of III–nitride μ-LEDs [61]. Figure 2.28 plots (a) the transient response of a μ-LED and a conventional broad-area LED and (b) the size dependence of the "turn-off" time, τ_{off}, of μ-LEDs. The turn-on response is on the order of our system response (~30 ps) and thus it cannot be measured. However, the turn-off transient is in the form of a single exponential and its life-time, τ_{off}, can thus be determined. It was found that τ_{off} decreases with decrease of μ-LED size. It reduced from 0.21 ns for *d* = 15 μm to 0.15 ns for *d* = 8 μm. This behavior is also expected since the effects of surface recombination are enhanced in smaller μ-LEDs. On the other hand, the increased operating speed may also be a result of an enhanced radiative recombination rate in μ-LEDs. With this fast speed and other advantages such as long operation lifetime, III–nitride μ-LED arrays may be used to replace lasers as inexpensive short-distance optical links such as between computer boards with a frequency up to 10 GHz.

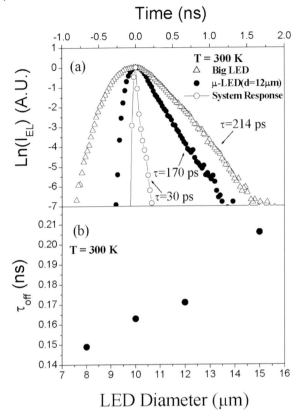

Fig. 2.28 (a) Transient responses of a μ-LED and a conventional broad-area LED. (b) Size dependence of the "turn-off" time, t_{off}, of μ-LEDs. After Ref. [61].

2.4.2
Vertical-Cavity Surface-Emitting LEDs

With the successful development of blue/green LEDs and purple LDs, vertical-cavity surface-emitting devices based on nitride semiconductors are currently also under investigation [62–64]. Although current-injected III–nitride VCSELs have not yet been fabricated, good progress has been made. For example, a current-injected vertical-cavity InGaN QW LED has been demonstrated recently by employing a combination of techniques [62–64], paving the way for future III–nitride VCSELs. A number of optoelectronic applications would benefit from VCSELs that feature a planar geometry and directional optical emission.

A current-injection vertical-cavity InGaN QW LED fabricated from an MOCVD-grown structure has been demonstrated by employing a combination of wafer bonding and substrate removal techniques. The nitride LED structures were separated from the sapphire substrate by a UV laser "lift-off" process, which allows the

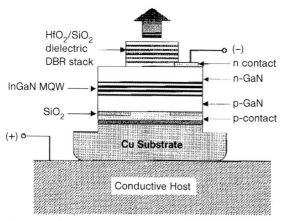

Fig. 2.29 Schematic drawing of a current-injection vertical-cavity LED with incorporation of a dielectric mirror stack. After Ref. [62].

transfer of the devices to a host substrate [62]. Patterned SiO_2 was used to define a current injecting aperture of 15–35 µm in diameter. The distributed Bragg reflector (DBR) was patterned and dry etched to define a comparable optical aperture size, with the electrical injection to p-type nitride provided by lateral current spreading via the indium tin oxide (ITO) layer. A HfO_2/SiO_2 multilayer DBR was deposited directly onto the exposed n-type nitride layer surface and patterned for completion of the optical resonator (consistent with the 15–35 µm effective aperture). A schematic drawing of the complete device is shown in Fig. 2.29.

Figure 2.30 displays the EL spectrum of a device at room temperature, at a current density of approximately $200\,A\,cm^{-2}$. Clear evidence of modal structure around the spectral peak at 460 nm has been observed, which is consistent with the estimated optical path of the resonant cavity. The emission at near 380 nm is interpreted as originating from other nitride layers outside the active InGaN MQW region. The EL spectrum presented in Fig. 2.30 clearly shows the high Q-value of the vertical cavity of this LED. To enhance the hole injection in the vertical cavity, a tunnel junction p^{++}/n^{++} InGaN/GaN bilayer [65] has been employed on the vertical-cavity LEDs and improved performance has been achieved [64]. The results suggested that it is feasible to design and construct higher-performance blue and near-UV vertical-cavity light emitters.

2.5
III–Nitride Microscale LED Applications

Although many of the ideas and potentials of these devices were identified long ago, it is only recently that a transition from basic research to practical device

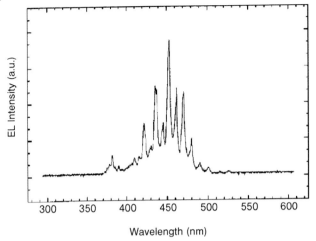

Fig. 2.30 The EL emission spectrum of a current-injection vertical-cavity LED at room temperature under DC operation, displaying cavity modes. After Ref. [62].

components has been made for microscale photonic devices due to various technological advances. Two such examples are discussed in this section, which include a novel light-emitting diode (LED) architecture based on interconnecting hundreds of microdisk LEDs together for enhancing LED efficiencies, and a microdisplay based on a III–nitride blue microscale LED array.

2.5.1
Interconnected Microdisk LEDs for Boosting LED Emission Efficiencies

Although broad-area LEDs possess the advantages of being simple to design, having low temperature sensitivity, and operating without threshold, they suffer from some limitations such as the poor extraction efficiency of light, the wide spectral width, and the large divergence of the light output. For many applications such as white-light LEDs for solid-state lighting, improvement of total efficiency or light output power is the most important issue. In this section, we discuss the method to replace conventional broad-area LEDs by hundreds of μ-LEDs to boost the LED power output. The novel architecture interconnects hundreds of microdisk LEDs (disk size on the order of 10 μm in diameter) made from InGaN/GaN single QWs. These interconnected microdisk LEDs fit into the same device area taken up by a conventional broad-area LED of $300 \times 300 \, \mu m^2$. It was shown that, for interconnected microdisk LEDs, the overall emission efficiency was increased by as much as 60% over the conventional LEDs for a fixed device area.

Interconnected μ-disk LEDs with two different individual μ-disk diameters were fabricated by photolithographic patterning and ICP dry etching [66]. Bilayers of Ni(20 nm)/Au(200 nm) and Al(300 nm)/Ti(20 nm) were deposited by electron-beam evaporation as p- and n-type ohmic contacts, respectively. The contacts were thermally annealed in nitrogen ambient at 650 °C for 5 min. An optical microscope image of fabricated interconnected InGaN/GaN QW μ-disk LEDs is shown in Fig. 2.31(a) side-by-side with that of a conventional broad-area LED. As shown in Fig. 2.31(a), more than a hundred μ-disks are interconnected and fit into an area of 300 μm × 300 μm. A transparent film of Ni/Au was deposited to connect the p-type ohmic contacts of each individual μ-LED in a net-like configuration. Though the fabrication of these interconnected μ-LED arrays requires different designs of masks than the conventional broad-area LEDs, the processing steps are identical to those of the conventional broad-area LEDs. It is thus expected that the manufactured yield of these novel LEDs will rival that of conventional LEDs.

The $I–V$ characteristics of interconnected μ-disk LEDs and a conventional broad-area LED fabricated from the same wafer were compared. The forward bias voltages at 20 mA, V_F, for the two types of LEDs are comparable and are about 3.7 V for 408 nm LEDs. In the earlier phase of our studies [66], V_F of the interconnected μ-disk LEDs was found to increase slightly over the conventional broad-area LED due to the reduction in the total active area as well as the total p-type contact area. However, later it was found that interconnected μ-disk LEDs provide improved current spreading and hence improved V_F values. The 300 K electroluminescence (EL) spectra of the interconnected μ-disk LEDs and broad-area LED have been measured. The EL spectral shapes of the interconnected μ-disk LEDs and the conventional broad-area LED are quite similar. However, more light is produced

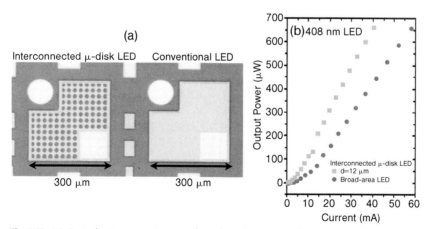

Fig. 2.31 (a) Optical microscope images of interconnected III–nitride μ-disk LEDs (left) and a conventional broad-area LED (right). (b) Comparison of the $L–I$ characteristics of an interconnected InGaN/GaN QW μ-disk LED with individual disk diameter of 12 μm and a conventional broad-area LED fabricated from the same wafer measured on the top surface of an unpackaged chip. The actual total output powers of these LEDs should be significantly higher than the measured values shown here since most of the light emission was not collected from these unpackaged chips. After Ref. [66].

by the interconnected microdisk LEDs at all experimental conditions, despite the fact that the total active area of an interconnected μ-disk LED is only about 60% of that of a broad-area LED of the same device area.

Figure 2.31(b) shows the light output power versus forward current for interconnected μ-disk LEDs of disk diameters of $d = 12\,\mu m$ and for a conventional broad-area LED measured on the top (p-type GaN side) of unpackaged chips. It should be pointed out that the actual total output powers of these LEDs should be significantly higher than the measured values shown here since most of the light emission was not collected from these unpackaged chips. The intriguing result shown in Fig. 2.31(b) is that the power output increases significantly in the interconnected μ-disk LEDs over the conventional broad-area LED, consistent with the EL measurement results. As illustrated in Fig. 2.31, we can achieve an overall quantum efficiency (QE) enhancement over 60% in interconnected μ-disk LEDs compared with the conventional broad-area LED.

The output power of LEDs, P_{output}, depends on the internal QE, η_{int}, and extraction efficiency, C_{ex},

$$P_{output} \propto \eta_{int} C_{ex} \tag{2.10}$$

An increase of P_{output} in the interconnected μ-disk LEDs is due to the enhancement of both the internal quantum efficiency (η_{int}) and extraction efficiency (C_{ex}). Additionally, the strain-induced piezoelectric field in the active QW regions may be reduced in microscale LEDs, resulting in increased internal quantum efficiency. However, we believe that the enhancement of extraction efficiency C_{ex} is probably more important here. The extraction efficiency C_{ex} of a conventional broad-area LED can be estimated by the following simple equation by considering the total internal reflection occurring at the LED/air interface. For GaN/air interface, $n_{GaN} \sin \theta_c = 1$, where $n_{GaN} = 2.4$, the critical angle θ_c is around 24°, providing an extraction efficiency around 8.3% from one side of the LED:

$$C_{ex} = \frac{\int_0^{2\pi} \int_0^{\theta_c} \sin\theta \, d\theta \, d\phi}{\int_0^{2\pi} \int_0^{\pi/2} \sin\theta \, d\theta \, d\phi} = (1 - \cos\theta_c) = 8.3\%$$

Low extraction efficiencies are one of the most serious problems for LEDs. As illustrated in Fig. 2.32, the emitted light is much easier to get out in the interconnected μ-disk LEDs than in the conventional broad-area LED. It is thus expected that C_{ex} is increased significantly in these interconnected μ-disk LEDs. Though the total active area is reduced, enhancements in both η_{int} and C_{ex} improved the overall QE of the interconnected μ-disk LEDs over the conventional broad-area LED.

We believe that the novel device could overcome two of the biggest problems facing LEDs: low extraction efficiencies due to the total internal reflection occurring at the LED/air interface, and the problem of current spreading. The present method of utilizing interconnected μ-disk LEDs for improving the brightness of LEDs is applicable to other semiconductor LEDs as well as polymer and organic

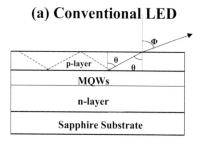

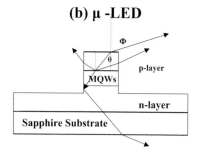

(a) Conventional LED

(b) μ -LED

$$R = \frac{\int_0^{2\pi} \int_0^{\theta_c} \sin\theta\, d\theta\, d\Phi}{\int_0^{2\pi} \int_0^{\pi/2} \sin\theta\, d\theta\, d\Phi} = (1 - \cos\theta_c) = 8.3\%$$

$$P_{output} \sim \zeta_{int} C_{ex}$$

ζ_{int} : Internal Quantum efficiency
C_{ex} : Extraction efficiency

Fig. 2.32 Schematic illustration of emission light extraction from (a) a conventional broad-area LED and (b) a μ-LED. Extraction efficiency is enhanced in interconnected μ-LEDs because emitted light is much easier to get out in the interconnected μ-disk LEDs.

LEDs. In particular, UV/blue LEDs based on III–nitrides are currently being used to generate white light by coating the chips with phosphors. In such an application, the use of the present method to generate more efficient UV/blue photons could be beneficial.

2.5.2
Nitride Microdisplays

Microdisplays are tiny displays with dimensions on the order of a few centimeters. However, they can provide a virtual image comparable to viewing a 21-inch diagonal TV/computer screen when put into eyeglasses and viewed through a lens system. Microdisplays can be used in a variety of devices such as head-worn and head-up displays, camcorders, viewfinders, etc., and have many commercial applications.

One such application of microdisplays is for head-up displays in vehicles. The drivers of modern vehicles are confronted with an increasing workload in terms of information processing. In order to meet the demand for a situation-adapted, nondistracting, rapidly perceptible information display, head-up displays can provide some advantages compared to instrument clusters. As illustrated in Fig. 2.33, in a driving situation using head-up display, the driver would see a virtual image reflected from the windshield into his/her eyes, which provides both less distraction from the road scene and less fatigue over driving periods and hence improved safety [67]. On the other hand, a head-up or head-worn microdisplay

Fig. 2.33 Head-up displays work by reflecting information on the windshield and superimposing it on the driver's view. The displays increase safety since drivers no longer have to take their eyes off the road. After Ref. [67], courtesy of Dr. Heinz-Bernhard Abel.

would provide a pilot not only with vital information, linking them to the aircraft's systems as well as to their rapidly changing environment, but also with hands-free capability, all of which could greatly enhance his/her ability to make split-second decisions and actions that coould determine the success or failure of a mission.

Current microdisplays are based on liquid-crystal display (LCD) technology [68] or organic light-emitting diodes (OLEDs) [69]. Semiconductor microdisplays, which require the integration of a dense array of microscale LEDs on a single semiconductor chip, have not previously been successfully fabricated. Furthermore, color conversion for full-color displays cannot be achieved in conventional III–V or Si semiconductors. So far, large flat panel displays based on semiconductor LEDs used on large buildings and stadiums have been made up of a massive number of discrete LEDs. However, III–nitride wide-bandgap semiconductors have a number of unique properties, including: variable bandgap from 0.8 eV (InN) through 3.4 eV (GaN) to 6.2 eV (AlN) with alloys; extremely high emission efficiency by incorporation of indium; high-power and high-temperature operation; extremely high shock resistance due to their mechanical hardness; and ease of color down-conversion from UV/blue/green to red or yellow. Due to these properties, there is no question that microdisplays fabricated from III–nitride wide-bandgap semiconductors can potentially provide superior display performance compared to LCD and OLED displays.

Table 2.1 summarizes some of the key features of microdisplays based on LCD, OLED, and the new III–nitride microdisplay technologies [70]. Microdisplays fabricated from III–nitrides can potentially provide superior display performance to LCD and OLED displays, thus providing an enhancement and benefit to the present capabilities of miniature display systems. Unlike LCDs that normally require an external light source, III–nitride blue microdisplays are self-luminescent and result in both space and power saving and allow viewing from any angle without color shift and degradation in contrast. On the other hand, current OLEDs must be driven at current densities many orders of magnitude lower than semiconductor LEDs to obtain reliable devices and hence are not suitable for high-intensity use. Additionally, III–nitrides are grown on sapphire substrates, which are trans-

Table 2.1 Comparison of III–nitrides with other technologies for microdisplays. After Ref. [69].

	Microdisplay technology		
	LCD	*OLED*	*III–nitride*
Luminance	<200 cd m^{-2} (full color) <2000 cd m^{-2} (green)	<1000 cd m^{-2} (green)	>1000 cd m^{-2} (full color) >10 000 cd m^{-2} (blue/green)
Contrast ratio	100 : 1 (intrinsic)	very high	very high
Resolution/pixel size	>10 μm	>10 μm	>6 μm
Response time	millisecond	microsecond	nanosecond
Color	full/mono	full/mono	full/mono
Viewing angle	>90°	>80°	>90°
Operating temperature	0 to +60 °C	−30 to +55 °C	−50 to +200 °C
Shock resistance	low	medium	high

parent to light and can hence serve as a natural surface for image display, reducing the steps for device packaging.

Blue microdisplays have been fabricated from InGaN/GaN QWs μ-disk LED arrays [71]. An optical microscope image of a prototype microdisplay (device dimensions of $0.5 \times 0.5 \, mm^2$) made up of 10×10 pixels of 12 μm diameter is shown in Fig. 2.34(a). To obtain a working device, a dielectric layer was deposited above the etch-exposed underneath n-type GaN layer to isolate the p-type contacts from the n-type layer. Shown in Fig. 2.34(a) are conducting wires that are used to make the connection between the n-type ohmic contacts and the contact pads, which are used for current injection into n-type ohmic contacts. There are also conducting wires that are used to make the connections between individual pixels through the top p-type ohmic contacts and the pixel control pads, which are used for current injection into p-type ohmic contacts. Each pixel has its own control pad. In this array, the state of the pixels can be individually controlled. The operation of these InGaN/GaN QW microdisplays has been demonstrated. Figure 2.34(b) shows optical microscope images of the blue microdisplay of Fig. 2.34(a) in action, displaying a sequence of letters: "KSU". The L–I characteristics of three individual 12 μm blue μ-disk LEDs within the microdisplay of Fig. 2.34(a) measured from the sapphire substrate side are shown Fig. 2.35, which demonstrated that the uniformity among these μ-disk LEDs is quite good. The inset of Fig. 2.35 shows an electroluminescence (EL) spectrum of blue μ-disk LEDs. The angular distribution of light emission from these μ-disk LEDs has been measured. The results demonstrated that microdisplays fabricated from III–nitride QWs could provide a very wide viewing angle.

In order to fabricate high-information-content matrix microdisplays, new concepts of bonding schemes must be created. For example, a real high-resolution microdisplay would contain 800×600 pixels or more. Since the sapphire substrate

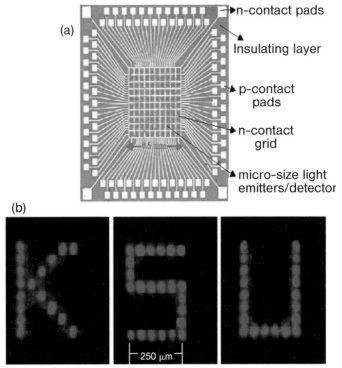

Fig. 2.34 (a) Optical microscopy image of a bonding scheme that allows one to address each microcavity pixel individually (or a III–nitride blue microdisplay). (b) Optical microscopy images (top view) of a III–nitride blue microdisplay in action (displaying the letters "KSU"), demonstrating the operation of the world's first semiconductor microdisplay made from III–nitride wide-bandgap semiconductors. After Ref. [71].

side is utilized to display images, flipchip bonding may be employed to integrate the μ-disk LED array with a matrix of discrete electrodes through the p-type contacts, each one corresponding to a point in the matrix. Full-color displays can also be realized by color down-conversion since InGaN/GaN QW μ-disk LEDs inherently emit blue light. Based on our preliminary results and the unique properties of III–nitrides, we believe that III–nitride microdisplays can potentially provide unsurpassable performance including: self-luminescence; high brightness, resolution, and contrast; high-temperature and high-power operation; high shock resistance; wide field of view; full color spectrum capability; long life; high speed; and low power consumption.

A 32 × 32 two-dimensional array of individually matrix-addressable blue InGaN QW LEDs has been fabricated [72, 73]. Each pixel of about 30 μm in diameter is equipped with its own integrated photoresist-based microlens (formed by a reflow technique) and can be electrically addressed with submillisecond access time. The

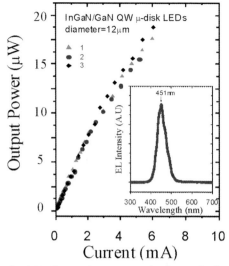

Fig. 2.35 The *L–I* characteristics for three individual microdisk LEDs within a microdisplay. The inset is an EL spectrum of a blue microdisk LED. All measurements were made through the sapphire substrate on unpackaged chips. The results show good uniformity of the microdisplay. After Ref. [71].

device fabrication steps are schematically illustrated in Fig. 2.36. In Fig. 2.36(a), following a two-step electron cyclotron resonance (ECR) plasma etch, the elements are defined and electrical contacts along rows and columns are made by standard (selective area) metallization to the n- and p-type regions, respectively. Figure 2.36(b) sketches an approach that utilizes planarization and a self-aligned technique in the process flow for array fabrication. Figure 2.36(c) is an optical microscopy image of a small group of devices around a particular element that has been switched on. The hemispherical microlenses made of photoresist covered each LED element, with a 50 µm pitch. The microlenses in this case had an approximately 30 µm focal length. These types of programmable microscale LED arrays are also useful as sources for proximity microscopy with high parallel throughput such as in spatially resolved fluorescence spectroscopy imaging applications.

It has been demonstrated by using an array of matrix-addressed micro-LEDs that it is possible to scale up to microdisplays containing a large number of elements [74, 75]. In this scheme, the underneath n-type layer of micro-LEDs in each column are connected in parallel, while the top p-layer of micro-LEDs in each row are connected in parallel. Each pixel (i.e., column i and row j) of the array can be turned on with a forward bias voltage applied between column i and row j. The array can be controlled by a line scan mechanism, which turns on the desired pixels in each row simultaneously and scans different row sequentially. A schematic diagram is shown in Fig. 2.37.

(a)

(b)

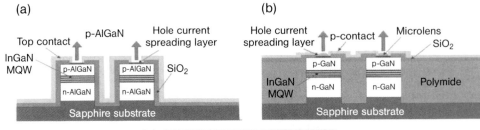

(c)

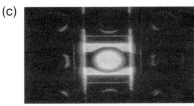

Fig. 2.36 Schematic diagram of fabrication step of a 32 × 32 two-dimensional array of individually matrix-addressable blue InGaN QW LEDs with pixel size of about 30 μm. (a) Following a two-step ECR plasma etch, the elements are defined and electrical contacts along rows and columns are made by standard (selective-area) metallization to the n- and p-type regions, respectively. (b) An approach that utilizes planarization and a self-aligned technique in the process flow for the array fabrication. (c) An optical microscopy image of a small group of devices around a particular element that has been switched on. After Refs. [72, 73].

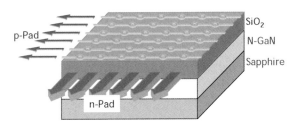

1. n-GaN column as (-) electrode
2. Ni/Au metal as (+) electrode
3. 2n(n=16 or 64) pads for n² array device

Fig. 2.37 Schematic diagram of a matrix-addressable micro-LED. After Refs. [74, 75], courtesy of Professor M. D. Dawson.

More recently, matrix-addressed InGaN/GaN QW micro-LED arrays with size up to 128 × 96 elements have been fabricated in matrix-addressed format [75]. The individual elements in the respective arrays are of diameter 20 μm and have a pitch of 30 μm, resulting in an overall chip size of about 2.5 mm × 3.5 mm for a 128 × 96 array. An optical microscopy image of a final fabricated microdisplay array of 64 × 64 elements is shown in Fig. 2.38.

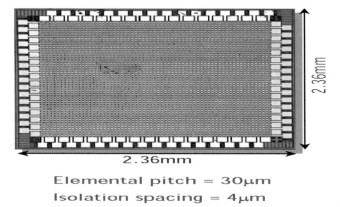

Elemental pitch = 30μm
Isolation spacing = 4μm

Fig. 2.38 Optical images of final devices: 64 × 64 matrix-addressable arrays of GaN-based μ-LEDs. After Refs. [74, 75], courtesy of Professor M. D. Dawson.

Since the fabrication of microscale LEDs is much less sophisticated than that for the vertical-cavity surface-emitting lasers (VCSELs), III–nitride blue micro-LED arrays are very attractive for inexpensive optical links. The capabilities of 2D array integration together with the advantages of high speed, high resolution, low temperature sensitivity, and applicability under versatile conditions make III–nitride μ-LEDs a potential candidate for light sources in short-distance optical communications. The bonding scheme of these prototype microdisplays can also be utilized to study the fundamental properties of individual microscale light emitters, including modified spontaneous emission, enhanced quantum efficiency, and lasing actions in microcavities, all of which are important for developing suitable materials and device designs for future III–nitride microscale optoelectronic devices such as micro-vertical-cavity LEDs, microcavity laser diodes, microscale detectors, and VCSELs.

2.5.3
III–Nitride Photonic Crystals

Photonic crystals (PCs) are artificially created ordered micro- and nanostructures in which media with different refractive indices are periodically arranged to create a photonic energy bandgap, a range of frequencies in which no electromagnetic modes can exist [76–79]. Similarly to the effect of the energy bandgap in semiconductors, the photonic bandgap dramatically changes the photon propagation properties. For example, if light is generated in a photonic crystal, it cannot propagate in a certain direction; if light is sent from outside, it will be completely reflected. Photonic crystals have attracted intense attention in scientific research and technical applications, because they offer approaches to manipulate photons, allowing the control of light propagation. These properties of PCs have been utilized in applications such as PC fibers for waveguiding and enhancing nonlinearity, and PC sensors. Of the many other potential practical applications, using PCs to dramatically increase the light extraction efficiency of LEDs is very promising. For LEDs, light extraction is a big concern. LED materials typically have a large refractive index, leading to a very small light extraction coefficient, with most generated light entrapped in the LED [76]. The enhancement of external efficiency of LEDs using PCs has been studied mainly in the infrared (IR) wavelength regions [76, 80–82], where the demonstrations have been carried out using optical pumping on semiconductor materials not yet fabricated into electrically pumped LED devices. A maximum of 30-fold enhancement in PL emission intensity from a free-standing InGaAs PC slab has been reported at a wavelength of 1100 nm [81]. A six-fold enhancement in emission intensity at 925 nm has been obtained from a LED structure upon PC formation under optical pumping [82].

Nitride PCs that function in the visible and UV regions have been studied only very recently due to the challenges associated with the fabrication of PCs with submicron periodicity [12–19]. In particular, the refractive index is about 2.4 for III–nitride LEDs, and the light extraction angle (or the critical angle for light to escape) is only ~23°, which amounts to only about 5–10% of the light extracted from the LEDs. The need for improvement in the extraction efficiency in LEDs is exceptionally great, especially for deep-UV LEDs ($\lambda < 340$ nm) based upon III–nitrides, which presently have very low quantum efficiency (QE). All lateral guided modes are lost due to parasitic absorption in UV LEDs. Chip-level innovation, such as the incorporation of PCs, is key to achieving a significant enhancement in QE of deep-UV photonic devices based upon III–nitrides. Several groups have exploited PCs to enhance the extraction efficiency of III–nitride LEDs [83–90].

Ideal photonic bandgap (PBG) is achieved by periodicity in three dimensions, but for extraction of light in LEDs, it is sufficient to eliminate light propagation only in the horizontal plane with the use of 2D PCs. For this purpose, a triangular lattice of air holes in a dielectric background are shown to be one of the most prominent 2D structures to present PBGs and are typically etched into semiconductor materials. The incorporation of PCs into III–nitride UV/blue emitters offers a solution for improving the extraction efficiency, but it is challenging. This is due in part to the difficulty in fabrication associated with the nanometer-scale features required for blue and UV wavelengths. Because PC structures must be built at the same scale as the wavelength to be suppressed, the difficulty of producing PCs increases with shorter wavelengths. Additionally, nitride photonic devices are generally grown on insulating sapphire substrates, which make electron-beam (e-beam) lithography patterning of nanoscale features difficult (due to charge accumulation). One other challenge is the ability to transfer the designed patterns from e-beam resist to the samples. Because the III–nitrides are hard materials, the most effective method to achieve this is by high-density plasma etching. This requires an optimized resist or a lift-off material that will withstand the plasma etching as well as maintain the small features of the PC.

III–nitride QWs and LED structures were grown by MOCVD for PC formation studies [84–87]. A high-quality AlN epilayer was inserted as a template, and a Si and Mg delta doping scheme was incorporated into the UV LED structures to reduce dislocation density and improve the n- and p-type conductivities. The active region was a single QW consisting of an AlInGaN quaternary or AlGaN ternary layer for UV LEDs. The active region was sandwiched between Si and Mg-doped AlGaN epilayers. Two-dimensional PCs have been successfully incorporated into both QWs and real LED devices with triangular lattice patterns of circular holes with varying diameter/periodicity down to 100/180 nm by e-beam lithography and inductively coupled plasma (ICP) dry etching.

To provide a reference, other sample devices did not receive the PC treatment [84–87]. Figure 2.39 shows the device layer structure and schematic design of the PC incorporation on a 330 nm UV LED and an SEM image of fabricated PCs on a UV LED. A hexagonal mesa of side length 120 μm, as shown in Fig. 2.39, was defined by e-beam lithography and etched by ICP dry etching. A hexagonal p-contact pad with 60 μm side length was deposited at the center of the LED mesa. To improve the electrical transport, a 10 μm wide n-type ohmic contact was deposited around the mesa along with a $100 \times 100 \, \mu m^2$ n-contact pad. Further details of the fabrication procedures are described elsewhere. PCs with triangular lattice patterns of circular holes with diameter varying from $d = 100$ nm to $d = 450$ nm and periodicity varying from $a = 300$ nm to $a = 700$ nm were fabricated on the LED everywhere (including the metal transparent layer) except the contact pads using e-beam lithography and ICP dry etching as described above. The I–V, L–I and electroluminescence (EL) measurements on specific LEDs were taken before and after the fabrication of PCs for comparison.

Optical characterization was first performed using near-field scanning optical microscopy (NSOM) uniquely configured for UV wavelengths [84]. The laser light was pumped on the unpatterned area of the sample about 10 μm outside the PCs and the emission intensity collected by the NSOM tip placed above the PCs. Emission intensity was also collected at an equal distance away from the same pump spot in an unpatterned region of the sample. The experimental setup for this measurement, as illustrated in the magnified inset of Fig. 2.40, would thus enable direct comparison between emission light intensity collected at the region pat-

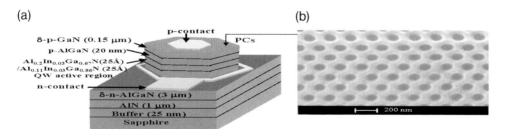

Fig. 2.39 (a) Schematic of photonic crystal (PC) incorporation on a III–nitride UV LED and (b) scanning electron microscopy (SEM) image of PCs created by nanofabrication on a UV LED. After Refs. [86, 87].

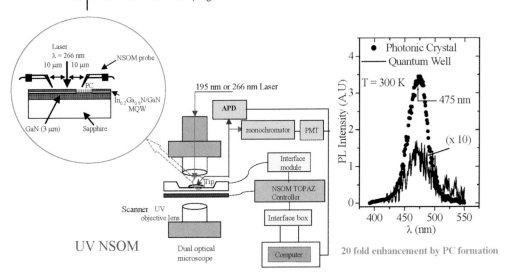

Fig. 2.40 Schematic setup of NSOM measurement of the light extraction by PCs. The NSOM system is uniquely configured for UV wavelengths (down to 200 nm). A magnified schematic of the sample is shown in the left circle. Shown on the right is the PL spectra collected from an InGaN/GaN QW sample at two regions (with and without PCs) that are equal distances away from the pump spot, as shown for the two cases above, measured by NSOM in collection mode. The results showed that we achieved a 20-fold enhancement of light extraction using optical pumping. After Ref. [84].

terned with PCs and an unpatterned region. The PL spectra measured for the two cases showed that the region patterned with PCs produced an emission intensity that was about 20 times higher than the unpatterned region of the sample. Additionally the NSOM results showed a 60° periodic variation with the angle between the propagation direction of emission light and the PCs lattice, and that the extraction of guided light traveling along the Γ–K direction can be as much as three times more than the light traveling along the Γ–M direction of the PCs in the nitride quantum well, which confirmed the effect of PCs.

Two-dimensional PCs with varying periodicity/diameter down to 150 nm/75 nm have also been successfully fabricated on AlN epilayers [91]. Deep-UV PL spectroscopy has been employed to study the optical properties of AlN PCs. With PC formation, a 20-fold enhancement in the band-edge emission intensity at 208 nm over unpatterned AlN epilayer was also observed. Because the formation of PCs reduces lateral guided modes, the result is a significant enhancement of light extraction in the vertical direction due to coupling into free space. Furthermore, the emission intensity increases with decreasing lattice constant of AlN PCs and the spectral peak energy decreases with decreasing lattice constant, indicating a possible release of compressive stresses as a result of PC formation. This implies that the PC feature is small enough to alter the strains in the AlN epilayers.

As guided by optical pumping studies, the Γ–K direction of the PCs was set perpendicular to the sides of the mesa to ensure a more efficient extraction of the guided light [86, 87], as schematically illustrated in Fig. 2.39. Figure 2.41 shows the optical microscopy images of 330 nm UV LEDs in action at a 20 mA injection current along with the power output versus current (L–I) characteristics. The less bright image in Fig. 2.41 is a conventional LED without PCs, while the brighter image is taken from the LED with PCs (or PC-LED), which clearly shows a significant enhancement of light output by PC formation. A typical electroluminescence (EL) spectrum of the 330 nm UV LEDs is shown in the inset of Fig. 2.41. No change was noticed in the peak position as well as the linewidth of the EL spectrum, indicating that the spontaneous emission is not significantly altered by the formation of PCs. The L–I characteristics of the 330 nm UV LEDs shown in Fig. 2.41 clearly demonstrate that the power output at 20 mA is about 2.5 times higher in the PC-LED than in the conventional LED without PCs. However, based on the optical pumping studies (which yielded an enhancement factor of 20), further improvements are possible by: (1) improving the conductivity of p-type AlGaN layers to allow a more effective current injection and spreading; (2) increasing etch depth of air holes to ensure penetration into the active layers; and (3) fabricating PCs with reduced diameter/periodicity to tune into the photonic bandgap.

In addition to the enhanced extraction of light from the LEDs, incorporation of PCs can provide an effective method to control the modulation speed of LEDs,

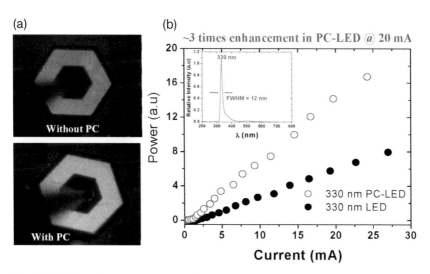

Fig. 2.41 (a) Optical microscopy images of 330 nm UV LEDs in action (top view). The output power enhancement is visible in the LED with PCs. (b) Comparison of L–I characteristics of 330 nm PC-LED with a conventional LED without PC. After Ref. [86].

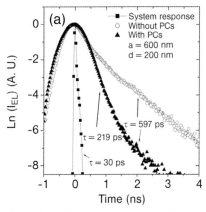
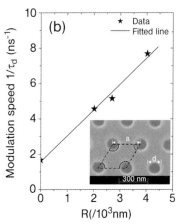

Fig. 2.42 (a) Transient responses of 330 nm UV LEDs with PCs ($a = 600$ nm, $d = 200$ nm) and without PCs. (b) The EL decay rate $(1/\tau)$ (or the relative modulation speed) plotted against R, where R is the etched sidewall perimeter per unit area, $R = 2\pi d/(3)^{1/2}a^2$. A linear fit of the data yields the sidewall surface recombination velocity. The inset is an SEM image of PCs along with the schematic to calculate R. The dashed parallelogram is the unit area and the curved lines at the four corners are etched perimeters. After Ref. [87].

which could be very useful for many applications [87]. With the incorporation of PCs on LEDs, the EL decay time constant decreases systematically with the increase of the etched sidewall area, attributed to the strong effect of surface recombination. Time-resolved EL results on 330 nm UV LEDs are shown in Fig. 2.42 and demonstrate that the incorporation of PCs provides four times enhancement in LED modulation speed. Based on these results, the surface recombination velocities on the sidewall of etched holes on 330 nm LEDs were determined to be $1.48 \times 10^5\,\mathrm{cm\,s^{-1}}$, which is comparable to values obtained by other methods [92] and is more than one order of magnitude smaller compared to the GaAs-based microstructures. This feature makes III–nitrides attractive for the fabrication of photonic structures with reduced dimensions.

2.6
Concluding Remarks

The aim of this chapter was to review the fabrication and optical studies of microscale structures and devices based on III–nitrides. With the exception of the III–nitride photonic crystal subsection, where a brief overview of the recent progress made in authors' laboratory was provided, it was our intention to cover articles published prior to June 2002, but we are sure that there may be related articles left out unintentionally. The III–nitride microstructures fabricated so far include microscale disk, ring, pyramid, prism, and waveguide structures. Active emitter devices achieved so far are microdisk cavity LEDs, vertical-cavity surface-emitting

LEDs and photonic crystal LEDs. The results obtained so far on III–nitride micro-cavity structures and devices are very promising, as demonstrated by several examples of practical device components discussed here. However, there are many problems and questions that still stand in the way of practical device implementation for many applications based on III–nitride microscale devices, such as micro-LED and detector arrays, VCSELs, as well as integrated photonics components. Achieving high quality and highly conductive materials, especially p-type $Al_x Ga_{1-x}N$ alloys with high Al contents, remains one of the foremost challenges of the nitride community. P-type conductive AlGaN alloys with relatively high Al contents are indispensable for hole injection for many photonics devices that are active in the deep-UV spectral region. At the same time, novel approaches for microscale ohmic contact fabrication, p-type contacts in particular, must be further developed to curtail the problem of relatively low p-type conductivity in III–nitrides. Additionally, due to the insulating nature of the sapphire substrates, packaging and bonding of these microdevices could present another challenge and innovative methods must be explored.

One of the potential applications for III–nitride microscale emitter arrays is for the sequencing of DNA. During the last 20 years, there have been remarkable advances in the use of fluorescence to study and map DNA [93]. It is now one of the most active fields in the life sciences. Most current technologies for DNA sequencing use laser-induced fluorescence for detecting the presence of a particular gene sequence. The principle of using fluorescence for DNA sequencing is that each DNA base (A, T, C, G) has its own unique characteristics under light excitation. The devices for DNA sequencing thus include a laser diode for excitation and a detection system to distinguish between different DNA bases. The current technology is cumbersome, expensive, and time-consuming: an array of DNA samples are prepared (microassay) and each sample is scanned one by one for DNA sequencing. Although the technology has advanced rapidly in the last 20 years, it still takes a long time to complete a sequence of DNA due to the vast number of DNA bases involved. Since III–nitride microemitters emit blue and UV light, which is the radiation used in DNA sequencing, it may be feasible to integrate III–nitride microemitter arrays with microassays of biological samples and micro-detector arrays, whereby it will be possible to map biological samples of microassays simultaneously as compared to current technology, which maps DNA sequentially. Since these arrays can, in principle, be fabricated on the scale of one million elements (1000 × 1000), the speed of DNA mapping can be increased by as many as six orders. Such a system is feasible with current existing technology. The real challenge is to integrate such a system together with microassays for biological and life science applications.

Because of the enormous potential for energy savings, as well as environmental benefits, there is currently a great interest worldwide in developing III–nitride LED technologies for solid-state lighting (SSL). However, significant improvement in the wall-plug efficiency has to be accomplished before the SSL mission could be achieved. Photonic crystal (PC) incorporation represents one of the most promising approaches. Another very interesting and promising approach is the incorporation of metal surface plasmons (SPs) [94]. In theory, with their high density of states, SPs can enhance the electron–hole recombination rate in QWs by coupling the emission into the SP mode of a thin metal film, and visible light can be efficiently extracted by SP mode scattering into air. Time-resolved PL measurements have confirmed that the recombination rate in GaN LED QWs can be significantly increased [95], and a 14-fold enhancement in peak PL intensity has been observed from silver-coated InGaN/GaN QW materials [96]. On the other hand, so far there has been no reported work on current-injected SP-coupled GaN LED devices. The fundamental difficulty involved in obtaining current-injected SP-coupled GaN LEDs originates from the III–nitride LED structural design. For example, the coupling spacing of SPs with excitons in QWs is limited to less than 50 nm, while for GaN LEDs this spacing is the p-GaN thickness, which has to be much more than 50 nm in order to achieve optimal performance. There are also technical difficulties related to the fabrication of both PCs and SPs. To achieve the optimal functions of PCs and SPs, patterned features will have a size in the sub-100 nm range, and there are currently no existing cost-effective approaches for the manufacturing of such nanostructures in the LED industry. Thus, future research will need to address manufacturing approaches to incorporate PC and SP technologies into GaN LED products for SSL applications.

Furthermore, active III–nitride photonics devices in the wavelength scale have not been fully explored and new physics are to be investigated. Nanophotonics involves photonics structures and devices with size comparable with the wavelength of the light they manipulate (λ/n), where n is the index of refraction, which is typically in the range of 100–400 nm. As the lateral size approaches the emission photon wavelength scale, the quantum nature of light is expected to play an important role with potentially an order-of-magnitude improvement in quantum efficiency. These nanophotonics devices range from arrays of nanolasers, nanodetectors, and nanowaveguides to optical switches and photonic crystals. Together with their two-dimensional array nature, the nanophotonics devices may open many important applications such as optical communications, signal and image processing, optical interconnects, computing, enhanced energy conversion and storage, and detection of chemical and biohazard substances and disease. In analogies to the importance of understanding the fundamental limit of electron transport in single-electron devices, the fundamental limit of light emitters and detectors must be understood before the potential of III–nitride nanophotonics becomes a reality. Smart self-assembling methods to grow and fabricate the nanostructures and nanodevices have to be sought. Fundamental issues including the size effects (and surface effects) on the optical and optoelectronic properties, carrier dynamics and optical resonance modes in submicron and nanoscale cavities must be fully explored. Recent work has shown that the surface recombination velocity in III–nitrides is an order of magnitude lower than those in other semiconductors [90, 91]. This, together with other unique features of III–nitrides, makes them very attractive for the study of nanophotonic structures and devices. It is our belief that III–nitride micro- and nanophotonics structures and devices will open many more important applications and is a burgeoning field with outstanding potential.

Acknowledgments

We are indebted to Dr. John Zavada, Dr. S. H. Wei, Professor R. Hui, and Professor M. D. Dawson for their valuable insights during the preparation of this review. We would like to acknowledge the contributions from some of our present and former group members: Dr. J. Li, Dr. Z. Y. Fan, Dr. T. N. Order, Dr. J. Shakya, Dr. K. H. Kim, Dr. K. B. Nam, Dr. M. L. Nakarmi, N. Nepal, Dr. R. Mair, Dr. K. C. Zeng, Dr. J. Z. Li, and Visiting Professor S. X. Jin. The research at Kansas State University was supported by grants from ARO, MDA, DARPA, NSF, DOE and ONR.

References

1 S. Nakamura and G. Fasol, *The Blue Laser Diode*, Springer, New York, 1997.

2 S. Nakamura, T. Mukai, and M. Senoh, *Appl. Phys. Lett.* 1994, 64, 1687.

3 M. Asif Khan, M. S. Shur, J. N. Kuznia, Q. Chen, J. Burn, and W. Schaff, *Appl. Phys. Lett.* 1995, 66, 1083.

4 H. Morkoc, S. Strite, G. B. Gao, M. E. Lin, B. Sverdlov, and M. Burns, *J. Appl. Phys.* 1994, 76, 1363.

5 S. Nakamura, M. Senoh, N. Iwasa, and S. Nagahama, *Jpn. J. Appl. Phys.* 1995, 34, L797.

6 "Nitride news," *Compound Semiconductor*, 1997, 3, 4.

7 H. X. Jiang and J. Y. Lin, "Advances in III–nitride micro-size light emitters," *III–Vs Review*, 2001, June/July issue, invited feature article.

8 R. K. Chang and A. J. Campillo, *Optical Processes in Microcavities*, World Scientific, Singapore, 1996.

9 Y. Yamamoto and R. E. Slusher, *Physics Today* 1993, 46, June, 66–73.

10 P. Gourley, *Scientific American* 1998, 278, 56.

11 M. Paul and S. Peercy, *Nature* 2000, 406, 1023.

12 T. Ito and S. Okazaki, *Nature* 2000, 406, 1027.

13 R. A. Mair, K. C. Zeng, J. Y. Lin, H. X. Jiang, B. Zhang, L. Dai, H. Tang, A. Botchkarev, W. Kim, and H. Morkoc, *Appl. Phys. Lett.* 1997, 71, 2898.

14 R. A. Mair, K. C. Zeng, J. Y. Lin, H. X. Jiang, B. Zhang, L. Dai, A. Botchkarev, W. Kim, H. Morkoc, and M. A. Khan, *Appl. Phys. Lett.* 1998, 72, 1530.

15 R. Mair, K. C. Zeng, J. Y. Lin, H. X. Jiang, B. Zhang, L. Dai, H. Tang, W. Kim, A. Botchkarev, H. Morkoc, and M. Asif Khan, *Proc. Mater. Res. Soc.* 1998, 482, 649.

16 J. C. Zolper and R. J. Shul, *Mater. Res. Soc. Bull.* 1997, Feb., 36.

17 K. C. Zeng, L. Dai, J. Y. Lin, and H. X. Jiang, *Appl. Phys. Lett.* 1999, 75, 2563.

18 J. W. Lee, C. B. Vartuli, C. R. Abernathy, J. D. Mackenzie, J. R. Mileham, R. J. Shul, J. C. Zolper, M. H. Crawford, J. M. Zavada, R. G. Wilson, and R. N. Schwartz, *J. Vac. Sci. Technol. B* 1996, 14(6), 3637.

19 C. Yoytsey, I. Adesida, and G. Bulman, *J. Electron. Mater.* 1998, 27, 282.

20 R. Underwood, D. Kapolnek, B. Keller, S. DenBaars, and U. Mishra, *Topical Workshop on Nitrides*, Nagoya, Japan, September 1995.

21 T. S. Zheleva, O. H. Nam, M. D. Bremser, and R. F. Davis, *Appl. Phys. Lett.* 1997, 71, 2472.

22 T. Akasaka, Y. Kobayashi, A. Ando, and N. Kobayashi, *Appl. Phys. Lett.* 1997, 71, 2196.

23 B. Beaumont, S. Haffouz, and P. Gibart, *Appl. Phys. Lett.* 1998, 72, 921.

24 S. Bidnyk, B. D. Little, Y. H. Cho, J. Karasinski, J. J. Song, W. Yang, and S. A. McPherson, *Appl. Phys. Lett.* **1998**, 73, 2242.

25 K. C. Zeng, J. Y. Lin, H. X. Jiang, and W. Yang, *Appl. Phys. Lett.* **1999**, 74, 1227.

26 K. Tachibana, T. Someya, A. Ishida, and Y. Arakawa, *Appl. Phys. Lett.* **2000**, 76, 3212.

27 T. N. Oder, J. Y. Lin, and H. X. Jiang, *Appl. Phys. Lett.* **2001**, 79, 12.

28 N. Oder, J. Y. Lin, and H. X. Jiang, *Appl. Phys. Lett.* **2001**, 79, 2511.

29 T. N. Oder, J. Li, J. Y. Lin, and H. X. Jiang, *Proc. SPIE* **2002**, 4643, 258.

30 http://www.phys.ksu. edu/area/GaNgroup

31 N. C. Frateschi, A. P. Kanjamala, and A. F. J. Levi, *Appl. Phys. Lett.* **1995**, 66, 1859.

32 N. C. Frateschi and A. F. J. Levi, *J. Appl. Phys.* **1996**, 80, 644.

33 R. P. Wang and M. M. Dumitrescu, *J. Appl. Phys.* **1997**, 81, 3391.

34 Lord Rayleigh, "The problem of the whispering gallery," *Scientific Papers*, Cambridge University, Cambridge, **1912**, Vol. 5, pp. 617–620.

35 S. Chang, N. B. Rex, R. K. Chang, G. Chong, and L. J. Guido, *Appl. Phys. Lett.* **1999**, 75, 166.

36 S. L. McCall, A. F. J. Levi, R. E. Slusher, S. J. Pearton, and R. A. Logan, *Appl. Phys. Lett.* **1992**, 60, 289.

37 N. C. Frateschi and A. F. J. Levi, *J. Appl. Phys.* **1996**, 80, 644.

38 Y. Kawabe, Ch. Spiegelberg, A. Schulzgen, M. F. Nabor, B. Kippelen, E. A. Mash, P. M. Allemand, M. Kuwata-Gonokami, K. Takeda, and N. Pryghambarian, *Appl. Phys. Lett.* **1998**, 72, 141.

39 B. C. Chung and M. Gershenzon, *J. Appl. Phys.* **1992**, 72, 651.

40 M. Smith, J. Y. Lin, H. X. Jiang, and M. A. Khan, *Appl. Phys. Lett.* **1997**, 71, 635.

41 G. D. Chen, M. Smith, J. Y. Lin, H. X. Jiang, S. H. Wei, M. Asif, and C. J. Sun, *Appl. Phys. Lett.* **1996**, 68, 2784.

42 J. A. Freitas, Jr., O. H. Nam, R. F. Davis, G. V. Saparin, and S. K. Obyden, *Appl. Phys. Lett.* **1998**, 72, 2990.

43 W. Li and W. X. Ni, *Appl. Phys. Lett.* **1996**, 68, 2705.

44 S. Chichibu, A. Shikanai, T. Azuhata, T. Sota, A. Kuramata, K. Horino, and S. Nakamura, *Appl. Phys. Lett.* **1998**, 68, 3766.

45 H. X. Jiang, J. Y. Lin, K. C. Zeng, and W. Yang, *Appl. Phys. Lett.* **1999**, 75, 763.

46 L. Dai, B. Zhang, J. Y. Lin, and H. X. Jiang, *Chin. Phys. Lett.* **2001**, 18, 437.

47 K. C. Zeng, M. Smith, J. Y. Lin, and H. X. Jiang, *Appl. Phys. Lett.* **1998**, 73, 1724.

48 K. C. Zeng, J. Li, J. Y. Lin, and H. X. Jiang, *Appl. Phys. Lett.* **2000**, 76, 3040.

49 P. Eckerlin and H. Kandler (eds.), *Numerical Data and Functional Relationships in Science and Technology*, Vol. III, Landolt-Bornstein, Springer, Berlin, **1971**.

50 Y. Masumoto, Y. Unuma, Y. Tanaka, and S. Shionoya, *J. Phys. Soc. Jpn.* **1979**, 47, 1884.

51 T. Katsuyama and K. Ogawa, *J. Appl. Phys.* **1994**, 75, 7607.

52 T. Katsuyama, S. Nishimura, K. Ogawa, and T. Sato, *Semicond. Sci. Technol.* **1993**, 8, 1226.

53 M. Matsuura and T. Kamizato, *Surf. Sci.* **1986**, 174, 183.

54 R. G. Ulbrich and G. W. Fehrenbach, *Phys. Rev. Lett.* **1994**, 43, 963.

55 D. E. Cooper and P. R. Newman, *Phys. Rev. B* **1989**, 39, 7431.

56 J. Y. Lin, Q. Zhu, D. Baum, and A. Honig, *Phys. Rev. B* **1989**, 40, 1385.

57 S. Nakamura, S. Senoh, S. Nagahama, N. Iwasa, T. Yamada, T. Matsushita, H. Kiyoku, Y. Sugimoto, T. Kozaki, H. Umemoto, M. Sano, and K. Chocho, *Appl. Phys. Lett.* **1998**, 72, 2014.

58 S. Nakamura, S. Senoh, S. Nagahama, N. Iwasa, T. Yamada, T. Matsushita, H. Kiyoku, Y. Sugimoto, T. Kozaki, H. Umemoto, M. Sano, and K. Chocho, *Jpn. J. Appl. Phys.* **1998**, 37, L1020.

59 T. Someya, R. Werner, A. Forchel, M. Catalano, R. Cingolani, and Y. Arakawa, *Science* **1999**, 285, 1905.

60 S. X. Jin, J. Li, J. Y. Lin, and H. X. Jiang, *Appl. Phys. Lett.* **2000**, 76, 631.

61 S. X. Jin, J. Li, J. Shakya, J. Y. Lin, and H. X. Jiang, *Appl. Phys. Lett.* **2001**, 78, 3532.

62 Y.-K. Song, M. Diagne, H. Zhou, A. V. Nurmikko, C. Carter-Coman, R. S. Kern, F. A. Kish, and M. R. Krames, *Appl. Phys. Lett.* **1999**, 74, 3720.

63 Y.-K. Song, M. Diagne, H. Zhou, A. V. Nurmikko, R. P. Schneider, Jr., and T. Takeuchi, *Appl. Phys. Lett.* **2000**, 77, 1744.

64 M. Diagne, M. Y. He, H. Zhou, E. Makarona, A. V. Nurmikko, J. Han, K. E. Waldrip, J. J. Figiel, T. Takeuchi, and M. R. Krames, *Appl. Phys. Lett.* **2001**, 79, 3720.

65 S.-R. Jeon, Y.-H. Song, H.-J. Jang, G. M. Yang, S. W. Hwang, and S. J. Son, *Appl. Phys. Lett.* **2001**, 78, 3265.

66 S. X. Jin, J. Li, J. Y. Lin, and H. X. Jiang, *Appl. Phys. Lett.* **2000**, 77, 3236.

67 http://w4.siemens.de/FuI/en/archiv/pof/heft1_02/artikel11/

68 N. Bergstrom, C. L. Chuang, M. Curley, A. Hildebrand, and Z. W. Li, *Society for Information Display Int. Symp., Digest of Technical Papers*, Vol. XXXI, May 16–18, **2000**.

69 J. Burtis, *Photonic Spectra*, October **2000**, p. 145.

70 H. X. Jiang and J. Y. Lin, *oe magazine (Monthly Publ. SPIE – Int. Soc. Opt. Eng.)*, **2001**, July issue, p. 28.

71 H. X. Jiang, S. X. Jin, J. Li, J. Shakya, and J. Y. Lin, *Appl. Phys. Lett.* **2001**, 78, 1303.

72 I. Ozden, M. Diagne, A. V. Nurmikko, J. Han, and T. Takeuchi, *Phys. Stat. Sol. (a)* **2001**, 188, 139 (Proc. 4th Int. Conf. on Nitride Semiconductors, Part A).

73 I. Ozden, M. Diagne, A. V. Nurmikko, J. Han, and T. Takeuchi, *Phys. Stat. Sol.(a)* **2001**, 188, 139.

74 C. W. Jeon, K. S. Kim, and M. D. Dawson, *Phys. Stat. Sol.(a)* **2002**, 192, 325.

75 C. W. Jeon, H. W. Choi, and M. D. Dawson, *Phys. Stat. Sol (a)* **2003**, 200, 79.

76 M. Boroditsky and E. Yablonovitch, *Proc. SPIE.* **1997**, 3002, 119.

77 E. Yablonovitch, *Phys. Rev. Lett.* **1987**, 58, 2059.

78 S. John, *Phys. Rev. Lett.* **1987**, 58, 2486.

79 J. D. Joannopoulos, R. D. Meade, and J. N. Winn, *Photonic Crystals*, Princeton University Press, Princeton, NJ, **1995**.

80 M. Boroditsky, R. Vrijen, T. F. Krauss, R. Coccioli, R. Bhat, and E. Yablonovitch, *J. Lightwave Technol.* **1999**, 17, 2096.

81 H. Y. Ryu, Y. H. Lee, R. L. Sellin, and D. Bimberg, *Appl. Phys. Lett.* **2001**, 79, 3573.

82 A. A. Erchak, D. J. Ripin, S. Fan, P. Rakich, J. D. Joannopoulos, E. P. Ippen, G. S. Petrich, and L. A. Kolodziejski, *Appl. Phys. Lett.* **2001**, 78, 563.

83 J. J. Wierer, M. R. Krames, J. E. Epler, N. F. Gardner, M. G. Craford, J. R. Wendt, J. A. Simmons, and M. M. Sigalas, *Appl. Phys. Lett.* **2004**, 84, 3885.

84 T. N. Oder, J. Shakya, J. Y. Lin, and H. X. Jiang, *Appl. Phys. Lett.* **2003**, 83, 1231.

85 T. N. Oder, H. S. Kim, J. Y. Lin, and H. X. Jiang, *Appl. Phys. Lett.* **2004**, 84, 466.

86 J. Shakya, K. H. Kim, J. Y. Lin, and H. X. Jiang, *Appl. Phys. Lett.* **2004**, 85, 142.

87 J. Shakya, J. Y. Lin, and H. X. Jiang, *Appl. Phys. Lett.* **2004**, 85, 2104.

88 L. Chen and A. V. Nurmikko, *Appl. Phys. Lett.* **2004**, 85, 3663.

89 A. David, C. Meier, R. Sharma, F. S. Diana, S. P. DenBaars, E. Hu, S. Nakamura, C. Weisbuch, and H. Benisty, *Appl. Phys. Lett.* **2005**, 87, 101107.

90 D.-H. Kim, C.-O. Cho, Y.-G. Roh, H. Jeon, Y. S. Park, J. Cho, J. S. Im, C. Sone, Y. Park, W. J. Choi, and Q.-H. Park, *Appl. Phys. Lett.* **2005**, 87, 203508.

91 N. Nepal, J. Shakya, M. L. Nakarmi, J. Y. Lin, and H. X. Jiang, *Appl. Phys. Lett.* **2006**, 88, 1331139.

92 M. Boroditsky, T. F. Krauss, R. Coccioli, R. Vrijen, R. Bhat, and E. Yablonovitch, *Appl. Phys. Lett.* **1999**, 75, 1036.

93 J. R. Lakowicz, *Principles of Fluorescence Spectroscopy*, 2nd edn., Kluwer, Dordrecht / Plenum, New York, **1999**.

94 R. H. Ritchie, *Phys. Rev.* **1957**, 106, 874.

95 A. Neogi, C.-W. Lee, H. O. Everitt, T. Kuroda, A. Tackeuchi, and E. Yablonovitch, *Phys. Rev. B* **2002**, 66, 153305.

96 K. Okanoto, I. Niki, A. Shvartser, Y. Narukawa, T. Mukai, and A. Scherer, *Nature Mater.* **2004**, 3, 601.

3
Nitride Emitters – Recent Progress

Tao Wang

3.1
Introduction

Since the pioneering works from Akasaki [1] in the late 1980s and Nakamura [2] in the early 1990s, GaN and its alloys have become established as the industry's preferred semiconductor materials for short-wavelength emitters over the last decade. Currently, the commercial use of semiconductor light emitters is progressing from long-wavelength light to short-wavelength light. The appearance of III–nitride emitters significantly changes the lives of humans in many respects. For example, solid-state lighting, mainly based on III–nitride materials, will result in a fundamental change in the concept of illumination experienced for more than 100 years. It will also massively save energy, estimated to be equivalent to \$112 billion by the year 2020 [3]. The market for III–nitride emitters can be divided into three classes: violet/blue/green light-emitting diodes (LEDs) emit between 400 and 540 nm; near-ultraviolet (UV) LEDs emit between 360 and 400 nm; and other UV LEDs emit below 360 nm. Currently, there are three approaches to achieve solid-state lighting: a package of three LED chips each emitting light of different wavelength (one red, one green, and one blue); a combination of a 460 nm LED with a yellow phosphor; and a single chip emitting UV light that is absorbed in the LED package by three phosphors (red, green, and blue) and re-emitted as broad-spectrum white light. In each of the above classes, the great majority of the demand is currently met by conventional light sources, i.e. incandescent filament lamps, fluorescent tubes, compact fluorescent tubes, and mercury lamps. However, LEDs are gaining market share across the whole range of wavelengths. The replacement of conventional light sources by LEDs will result in two key benefits: more than 10 times higher efficiency of conversion of electricity to light, and more than 10 times longer lifetime. Other benefits include much faster response and resistance to mechanical shock in some particular applications.

There are important applications of UV light sources in many fields, such as environmental protection, medical devices, as well as illumination. For example, one of the strongest air pollution sources is from oil-based engines. Air purifiers

Wide Bandgap Light Emitting Materials and Devices. Edited by G. F. Neumark, I. L. Kuskovsky, and H. Jiang
Copyright © 2007 WILEY-VCH Verlag GmbH & Co. KGaA, Weinheim
ISBN: 978-3-527-40331-8

can decompose organic substances in the air into water and carbonic acid using titanium oxide as an optical catalyst, which works well only under the illumination of UV light. In fact, 380 nm LEDs have been demonstrated for this application, but their efficiency is very low. Using 350 nm LEDs instead, the efficiency of titanium oxide as the optical catalyst for decomposing organic substances is significantly improved by a factor of 8. UV light can be used in curing a wide range of coatings (including glues, inks, and laminates) by catalyzing polymerization to increase viscosity. Absorption of deep-UV light with wavelengths of 280–250 nm disrupts bacteria's DNA, rendering it harmless. The optimum wavelength for this disruption in an aqueous environment is 260 nm. The first experimental demonstration of disinfection using a 280 nm GaN LED was reported in May 2004 [4]. More recently, using a 280 nm UV LED with an output power of 7.34 mW, the level of *Escherichia coli* in spiked water flowing at 150 ml min^{-1} and 300 ml min^{-1} were reported [5] to be reduced by 99.99% and 99.0%, respectively, close to values required for individual water treatment systems, typically at a rate of 500 ml min^{-1}. The first commercial application of LED sterilization is anticipated to be portable equipment for defense use. The market for UV sterilization is forecast to grow substantially as environmental standards rise. It is already used for all bottled water. US legislation is increasingly requiring its use on tap water. UV sterilization is also growing for wastewater, as an environmentally preferable alternative to using chlorine.

The global market for all high-brightness LEDs will be $5 billion in 2006, and it will rapidly grow to $10.8 billion for different applications in 2010 [6], among which general illumination will become the fastest-growing application between now and 2010. By the year 2010, the high-brightness LED market for general illumination is forecast to be $3.2 billion [6]. In the global UV light source market, UV LEDs currently hold a negligible share of the market [7], but this will increase remarkably with the development of new technologies for deep-UV LEDs with emission wavelengths down to 250 nm. Recently, Hu et al. [8] reported a 1.2 mW, 280 nm LED under a 20 mA injection current, and Allerman et al. [9] demonstrated a 237 nm LED, the shortest-wavelength LED so far reported. This implies that it is not too far to replace the currently used mercury lamp by solid, compact, and efficient UV LEDs.

Short-wavelength lasers are currently limited to specialist applications, their sales of $9 million representing less than 0.5% by value of all laser diodes [10]. However, the short-wavelength laser has been adopted as the laser for next-generation DVDs and their sales are therefore forecast to grow to $272 million by 2008 [10]. This market is dependent upon the uptake of next-generation DVDs, e.g. for use with next-generation high-definition TV.

With "violet/blue" technology on polar (0001) sapphire or SiC substrates approaching maturity, current research on III–nitride optical devices tends to be directed toward UV emitters, high-power green emitters, UV-to-visible emitters grown on silicon substrates or along the nonpolar direction, and white LEDs using a single chip without involving any phosphor. This chapter will discuss recent progress on developing III–nitride optical devices in these areas. Section 3.2

describes the progress in the area of UV LEDs in two classes, one based on GaN buffer technology for UV emitters in the UV-A spectral region in Section 3.2.1, and another based on GaN-free technology for the deep-UV emitters mainly in the UV-B and UV-C spectral regions in Section 3.2.2. The various technological approaches that have led to remarkable achievements in UV LEDs are described in detail. The good progress in the area of UV lasers is presented, and alternative concepts to achieve low-threshold lasers are also discussed in Section 3.2.3. Section 3.3 describes the research efforts on InGaN-based green LEDs on sapphire and InGaN-based LEDs grown on silicon substrates. Currently, the performance of single-chip white LEDs that do not involve any phosphor is far from satisfactory. Although there has not been a breakthrough in growth technology, some developments have been made, which are included in Section 3.3. Recently, in order to eventually eliminate polarization-induced internal electric fields, there tends to be more and more interest in III–nitride emitters grown along nonpolar or semipolar directions, which is particularly important for long-wavelength III–nitride LEDs, like green LEDs. Recent advances in this area are discussed in Section 3.4. Finally, Section 5 gives a summary of this chapter and discusses perspectives on further improvements of III–nitride optical devices.

3.2
Ultraviolet Emitters

UV light covers a wide region in the spectrum. The wavelength is from 100 to ~400 nm. Generally speaking, UV light is arbitrarily broken down into three bands, namely, UV-A (315–400 nm), UV-B (280–315 nm), and UV-C (100–280 nm). In theory, the band emission of Al-containing III–nitride materials can be extended down to ~200 nm, covering a major part of the UV spectrum. Currently, the predominant UV light source is the mercury vapor lamp, in two classes. A narrow single-emission peak centered at 254 nm but a relatively low density of output power can be obtained by a low-pressure mercury lamp; while a medium-pressure lamp emits a higher output density but multiple-peak spectrum in a wide range from 250 to 370 nm, with a highest intensity peak at 365 nm. A high-voltage arc is normally required to initiate discharge, and the lifetime is limited to 1000 h. There are some other serious problems in limiting the application of the mercury vapor lamp. Obviously, it is not suitable for portable use, easily causes environmental pollution, and lacks tenability in wavelength. Other disadvantages include a slow speed of response and large energy consumption. Therefore, solid and compact UV LEDs with good tunable wavelengths will eventually replace mercury vapor lamps.

Since the early 1990s, there have been major achievements in developing InGaN/GaN multiple-quantum-well (MQW) LEDs with high optical power and InGaN/GaN-based laser diodes (LDs). Currently, InGaN/GaN-based violet/blue LEDs with an output power of 10–20 mW at a 20 mA injection current have been commercialized, and continuous-wave (CW) LDs with a lifetime of 10 000 h have been used for high-density storage DVDs. However, when the emission wavelength decreases to below 370 nm, the optical quantum efficiency dramatically reduces as the indium mole fraction falls to almost zero [11], which is believed to be mainly due to the disappearance of the so-called exciton localization effect. Below 362 nm (i.e., GaN band-edge emission), InGaN can be difficult to use as an active region unless an Al-containing layer with a high Al composition is used as a potential barrier for quantum wells. This may lead to particular difficulty as the emission shifts toward shorter wavelength. Instead, the Al-containing III–nitride material has to be used as an emitting region. Compare to the InGaN-based violet/blue/green emitters, the performance of the current UV emitters is quite poor due to a number of reasons, including poor crystal quality, cracking issues, and large activation energy for both p- and n-AlGaN, in particular, for high Al-containing layers.

The early reports included a 375 nm UV LED based on double heterostructure (DH) AlGaN/GaN in 1994 [12], and a DH AlGaN/GaN injection diode with an observed weak peak at 360 nm in 1995 [13]. A 350 nm UV LED, based on a GaN/AlGaN quantum well as an active region, was reported to show an output power of 13 μW at 20 mA injection current in 1998 [14]. Recently, there were some significant improvements in UV LEDs in terms of emission wavelength and optical power. More recently, with the development of AlN buffer technology, a great achievement has been made, pushing the emission wavelength of UV LEDs down to below 250 nm [8, 15].

3.2.1
Ultraviolet LEDs Based on GaN Buffer Technology

For LEDs in the UV-A spectral region, the GaN buffer technology is still reasonably good to achieve UV LEDs in spite of the existence of an internal absorption issue, since the crystal quality of the GaN layer on sapphire is generally higher than that of the AlN layer in terms of dislocation density. Second, LEDs in the UV-A spectral region do not require a high Al composition. Therefore, the lattice mismatch between GaN and AlGaN is not so large, which does not cause a significant degradation in quality. However, the risk of formation of cracks should be carefully avoided. There are a few approaches to realize UV LEDs in the UV-A spectral region. The reports in the early 2000s [16–21] were mainly based on the GaN buffer technology on sapphire [16, 19–21] and a few on SiC substrate [17, 18]. Otsuka et al. [16] reported a UV LED with an emission wavelength of 339 nm on sapphire in 2000, in which a DH $Al_{0.13}Ga_{0.87}N/Al_{0.10}Ga_{0.90}N$ was used as an active region. It has been realized that the crystal quality of the GaN buffer becomes a critical issue in making high-performance UV LEDs, unlike InGaN-based LEDs. Along with more efforts on improving crystal quality using different approaches, some groups attempted to use GaN substrates. Using GaN as a substrate, the crystal quality can be greatly improved, and the optical power of a 352 nm UV LED on GaN substrate was greatly increased to 0.55 mW at 20 mA injection current, measured based on a bare chip, as reported by Nishida et al. in 2001 [19]. In order to suppress the internal polarization field, an AlGaN-based quantum well as an active region was used instead of a GaN quantum well. The active region consisted of a 2.5 nm

$Al_{0.04}Ga_{0.96}N$ single quantum well, sandwiched by $Al_{0.1}Ga_{0.9}N$ barriers. A short-period superlattice as a cladding p-layer was used to improve the transparency and conductivity. A maximum power exceeding 10 mW at 400 mA current and an internal quantum efficiency of more than 80% indicated that a highly efficient UV LED could possibly be achieved by careful device design (light extraction, internal absorption, and strain management) and improving crystal quality. Also, the *I–V* curve showed 3.8 V and 4.6 V at 20 mA and 100 mA injection current, respectively. Both the *I–V* and *I–L* characteristics meant that there should be chances to potentially achieve UV LEDs with performances that are comparable to InGaN-based LEDs. Based on a modified epitaxial lateral overgrowth (ELOG) technique, Akasaki's group reported a 0.6 mW, 352 nm LED on sapphire substrate at 20 mA injection current, and a 3.2 mW, 363 nm UV LED at 100 mA injection current in 2001 and 2002, respectively [20, 21]. Both were measured in CW mode.

The ELOG and evolved technologies have proved to be an efficient way to yield a significant reduction of dislocations in GaN films. However, these approaches heavily involve ex-situ patterning processes. In parallel to further developing the ELOG technology, more efforts have been put in to developing in-situ mask patterning technologies. Wang et al. [22] reported an in-situ Si_xN_y mask patterning approach to significantly improve the crystal quality of GaN grown on sapphire. At a low temperature (500 °C), a thin Si_xN_y layer was initially deposited on sapphire with an optimum thickness, which was completely different from the previous report on the Si/N treatment at high temperature. After this step, the growth was carried out in a standard metal–organic chemical vapor deposition (MOCVD) growth mode for GaN, i.e. the growth of a low-temperature GaN buffer layer was followed by a nominally undoped GaN layer grown at high temperature. Transmission electron microscopy (TEM) indicated a significant reduction in dislocation density, compared with GaN grown using the standard two-step technique. Figure 3.1 shows a TEM cross-sectional image of the GaN grown using the in-situ Si_xN_y mask, indicating that the high dislocation density often appearing in conventional GaN disappears. Further evidence included X-ray diffraction (XRD) data, showing a reduction in width of the XRD rocking curve. The same group [23] reported a highly improved optical performance of a 350 nm UV LED using the in-situ SiN mask technology. Their structure was prepared as follows. Following the growth of the thin Si_xN_y in-situ mask and a standard thin GaN nucleation layer, a 2 μm thick Si-doped GaN layer was grown. Then 500 periods of a Si-doped AlGaN(2 nm)/GaN(2 nm) strained-layer superlattice were used as the bottom cladding layer. The active region consisted of a 2 nm GaN single quantum well (SQW) separated by 10 nm AlGaN barriers. Finally, 50 periods of Mg-doped AlGaN(2 nm)/GaN(1 nm) superlattice (the top cladding layer) and then a 20 nm thick Mg-doped GaN layer were grown. The *I–L* curve, measured from a bare chip with a standard size of $350 \times 350 \, \mu m^2$, showed an improvement in optical power by a factor of 2. However, when the current was higher than 120 mA, the optical power was suddenly decreased, which might result from a serious heating effect due to the introduction of the in-situ Si_xN_y mask.

High-performance InGaN-based violet/blue/green LEDs on sapphire have been realized in terms of optical power in spite of a high dislocation density. This

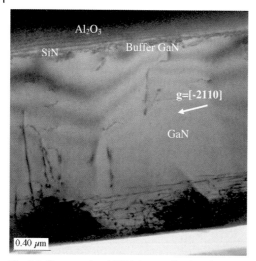

Fig. 3.1 Cross-sectional TEM image of a GaN layer using a thin Si_xN_y layer as a mask deposited at low temperature, showing that the high dislocation density often appearing in conventional GaN has disappeared. From Ref. [22].

insensitivity of efficiency to dislocation density has generally been accepted as being due to the so-called exciton localization effect, caused by the nonuniform indium composition in an InGaN active layer. The degree of carrier localization decreases with decreasing indium content. Therefore, it is necessary to increase the degree of the exciton-localization effect in the active layer and to decrease nonradiative recombination centers to improve the quantum efficiency at short wavelength. There are a few methods to produce exciton localization, by modifying either the thickness or the composition of the quantum well through the introduction of indium to form an AlInGaN quaternary system. Lee et al. [24] reported a Ga-droplet approach, by which a fluctuation of GaN/AlGaN quantum-well thickness was produced. By means of only flowing the Ga precursor [trimethylgallium (TMGa)] for 2 s, a Ga droplet layer was grown prior to growth of an AlGaN layer, which was used as a barrier for the quantum well. The AlGaN(20 nm):barrier/GaN(2 nm):SQW/AlGaN(20 nm):barrier as the active layer was then grown. Finally, 50 periods of Mg-doped AlGaN(1 nm)/GaN(1 nm) superlattice and then a 20 nm thick Mg-doped GaN layer were grown. The role of following TMGa is to form Ga droplets, which are expected to modify the well thickness in order to produce exciton localization. A detailed TEM study indicates that clear contrasts due to the formation of Ga droplets are formed, as marked by arrows in Fig. 3.2. The further measurement of the *I–L* curve, based on a chip with a standard size of 350 × 350 μm², indicated that the optical power was improved by a factor of ~2, compared with a similar structure but without using the Ga-droplet approach.

Using an AlInGaN quaternary is another approach to produce exciton localization, which is highly expected to improve quantum efficiency. It is generally

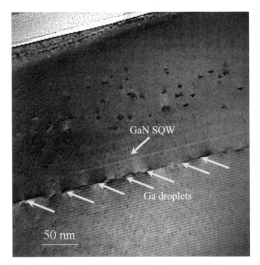

Fig. 3.2 Cross-sectional TEM image of a UV LED grown using the Ga-droplet approach, showing clear contrasts at the interface, shown by arrows. From Ref. [24].

believed that using the AlInGaN quaternary as an active region is one of the best solutions to achieve high-performance UV LEDs on large lattice-mismatched sapphire substrates even without the involvement of the ELOG technique. Zhang et al. [25] reported a 0.12 mW, 343 nm UV LED using the AlInGaN quaternary material as the active region grown by a pulsed atomic layer epitaxial technique (PALE), basically evolved from the atomic layer epitaxial technique. This technology will be discussed later. In their structure, four periods of $Al_{0.36}In_{0.01}Ga_{0.63}N/Al_{0.2}In_{0.03}Ga_{0.77}N$ MQWs were used as the active region. One of the main advantages of PALE is that the migration of Al and Ga adatoms can be enhanced at the low temperature required for growth of indium-containing alloys. Wang et al. [26] reported a 348 nm UV LED with a significantly improved optical performance by carefully designing the device structure. The active region consisted of an AlInGaN SQW separated by AlGaN barriers, where the Al mole faction was about 20% and 6% for the barrier layer and the well layer, respectively, and the indium mole fraction was around 3.5%. The barrier thickness was 20 nm. In order to decrease the piezoelectric field, normally resulting in the degradation of optical performance, a thin well thickness was intentionally designed. Figure 3.3(a) shows the *I–L* and *I–V* characteristics and electroluminescence (EL) spectrum of an AlInGaN-based UV LED. The optical power is 0.3 mW at 20 mA injection current, and it increases to 1 mW at 50 mA. Both are improved by one order magnitude compared with a UV LED using a GaN/AlGaN quantum well as an active region. Figure 3.3(b) shows a narrow EL peak at 348 nm under 20 mA injection current. The evidence to prove the existence of strong exciton localization includes the temperature-dependent photoluminescence measurement shown in Fig. 3.4. The emission

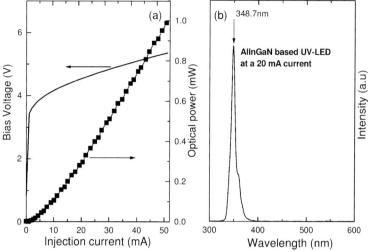

Fig. 3.3 (a) The *I–L* and *I–V* characteristics of an AlInGaN-based UV LED. The optical power is 0.3 mW at 20 mA injection current, and it increases to 1 mW at 50 mA. (b) Electroluminescence spectrum at 20 mA, where a strong emission at 348 nm appears. From Ref. [26].

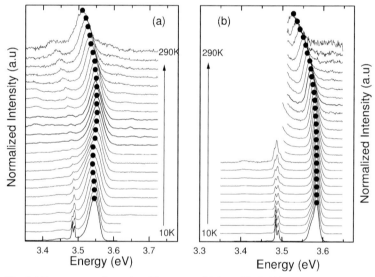

Fig. 3.4 Temperature-dependent PL spectra of (a) an AlInGaN/AlGaN quantum well, showing a clear "S" shape of peak emission energy, and (b) an GaN/AlGaN SQW, showing a monotonic decrease in peak emission energy. In both cases, the peaks on the low-energy side are from the GaN buffer layer underneath. The dots are guides for the eye only. From Ref. [26].

energy shows a clear blueshift between 70 K and 130 K for the AlInGaN quaternary in Fig. 3.4(a), a typical fingerprint for exciton localization, often observed in the InGaN/GaN system. In contrast with this, there is no clear blueshift to be observed in the GaN/AlGaN quantum well, as shown in Fig. 3.4(b).

Recently, Hirayama et al. [27] further improved the performance of the AlInGaN-based 350 nm UV LED using GaN substrate, leading to a 7.4 mW optical power under 400 mA injection current, which is superior to the AlGaN-based 350 nm LEDs on GaN substrate, previously reported by Nishida et al. [19]. The maximum external quantum efficiency is ~1% at 50 mA injection current.

One of the factors limiting the optical performance of UV LEDs using the GaN buffer is due to internal absorption, which has to be solved. Obviously, the incorporation of distributed Bragg reflectors (DBRs) below the light-emitting region is a good option. Also, DBRs have been demonstrated to greatly improve the optical extraction efficiency. Recently, high-quality crack-free DBRs with a peak reflectivity of 91% at ~350 nm grown by MOCVD have been reported by Wang et al. [28], as shown in Fig. 3.5. The GaN buffer technology was used to initially grow a 1 μm thick layer of nominally undoped GaN, followed by a 30 nm AlN interlayer to prevent the formation of cracks during the subsequent growth of 25 pairs of quarter-wave DBRs, consisting of 36.5 nm $Al_{0.49}Ga_{0.51}N$ and 34.5 nm $Al_{0.16}Ga_{0.84}N$ in each pair. The cross-sectional scanning electron microscopy image shown in Fig. 3.5(b) indicates the 25 pairs of DBRs clearly, from which the average bilayer thickness is estimated to be 71 nm. Figure 3.5(a) shows the room-temperature (RT) reflectivity spectrum of the DBRs, exhibiting a symmetric stopband with a flat-topped reflectivity peak. The peak reflectivity at 353 nm is 91%, and the stopband width is around 17 nm, both indicating a high quality of this sample. With further modifying the Al compositions of individual layers in each pair, 94% reflectivity from 25 pairs of $Al_{0.60}Ga_{0.40}N/Al_{0.20}Ga_{0.80}N$ DBRs was obtained [29]. Following the successful growth of the high-reflectivity DBRs, a DBR-enhanced UV LED with an emission wavelength of 350 nm was reported by the same group [30]. The 25 pairs of quarter-wave DBRs were grown prior to the growth of a 90-period n-type $Al_{0.25}Ga_{0.75}N(2\,nm)/GaN(2\,nm)$ superlattice as the bottom cladding layer. The DBRs are simply used as a refractor to avoid internal absorption. The active region consisted of 10 periods of $Al_{0.05}In_{0.045}Ga_{0.905}$ $N(2.5\,nm)$:well/$Al_{0.2}In_{0.002}Ga_{0.78}N(5.0\,nm)$:barrier MQWs, sandwiched by both the bottom cladding layer and the 50-period p-type $Al_{0.25}Ga_{0.75}N(2\,nm)/GaN(1\,nm)$ superlattice as the top cladding layer. Finally, a 10 nm p-GaN is capped for a good p-contact. Figure 3.6 shows the output power as a function of injection current, indicating that the output power is increased by a factor of ~2.3. This value agrees well with the calculation based on a photo-recycling model. This is the first DBR-enhanced UV LED reported so far. Using a lattice-matched system like $Al_{0.85}In_{0.15}N/$ $Al_{0.2}Ga_{0.8}N$ aiming at the UV spectral region, a further improvement has been achieved. Feltin et al. [31] reported ~99% peak reflectivity at 363 nm using 35 pairs of lattice-mismatched $Al_{0.85}In_{0.15}N/Al_{0.2}Ga_{0.8}N$ DBRs. The 35 pairs of DBRs were grown on 1 μm $Al_{0.2}Ga_{0.8}N$ on a standard GaN buffer on sapphire. In order to avoid the formation of cracks, a thin GaN/AlN superlattice was applied.

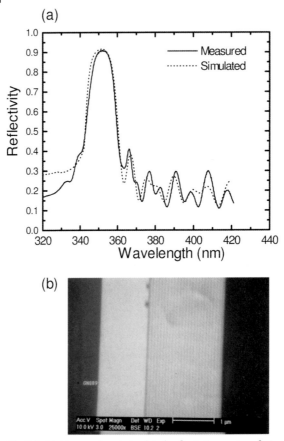

Fig. 3.5 (a) Measured and simulated reflectivity spectra of 25 pairs of DBRs, showing a symmetric flat-topped stopband. The peak reflectivity at ~350 nm is 91%. (b) Cross-sectional SEM image of the 25-pair DBR, where the average bilayer thickness is estimated to be 71 nm. From Ref. [28].

However, this lattice-matched DBR has not yet been applied in the growth of UV emitters.

3.2.2
Ultraviolet LEDs Based on GaN-Free Technology

However, it may become difficult to carry on using the GaN buffer technology to achieve LEDs in the UV-B and UV-C spectral regions due to a number of limiting factors, including the serious crack issue. The further increase in lattice mismatch between GaN and AlGaN leads to a significant degradation in crystal quality. In this case, an AlGaN layer should be used as a buffer grown at a high temperature

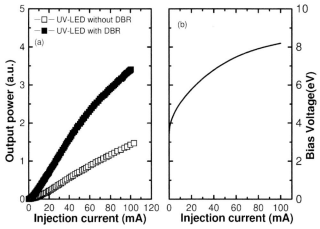

Fig. 3.6 (a) Output power of a UV LED with and without DBRs as a function of injection current up to 100 mA in CW operation mode. The optical power with DBRs is increased by a factor of 2.3, compared with the UV LED without DBRs. (b) The *I–V* characteristic. From Ref. [30].

after a thin AlN nucleation layer deposited at a low temperature, namely, the two-step technique. With further increasing Al composition in the AlGaN buffer layer, Khan's group [32, 33] used an AlInGaN quaternary as an active region to push the emission wavelength of a UV LED down to 315 nm. By modifying the compositions of Al and In in the active region, the emission wavelength of the LEDs further shifted to 305 nm. Hirayama and coworker also carried out some nice work on AlInGaN-based LEDs, summarized in Ref. [34]. They used an alternating gas flow growth method similar to the PALE mentioned above, leading to an improvement in the crystal quality of p-AlGaN with a high Al composition. They also reported 308 and 314 nm UV LEDs on sapphire substrates using thick $Al_{0.47}Ga_{0.53}N$ as a buffer layer, respectively. The maximum optical power was 0.4 mW at 130 mA injection current for 308 nm LEDs, and 0.8 mW at 260 mA injection current for 314 nm LEDs. In their design, a 60 nm $Al_{0.46}In_{0.02}Ga_{0.42}N$ single layer was used as the emitting region, basically a DH injection diode. So far, the 305 nm LED may be the LED with the shortest emission wavelength using the AlInGaN quaternary system as the emitting region. It might be difficult to use this quaternary system to further shift toward shorter wavelength since the growth of high Al-containing alloys generally requires a high temperature, which may result in the difficulty in incorporating indium into the AlGaN layer.

In order to move toward shorter wavelength, AlN as a buffer layer may become critically important. Early research on the growth of AlN on sapphire was aimed at obtaining a pit-free surface. Ohba and Hatano [35] studied the influence of temperature and NH_3 flow rate on growth rate, indicating that high growth temperature and low V/III ratio were key parameters. A low V/III ratio is helpful for reduction of the pre-reaction between trimethylaluminum (TMAl) and NH_3, leading to a higher growth rate of AlN. In the year 2000, the same group [36] developed a two-step growth technique. First, a thin AlN layer (20 nm) was grown at 1200 °C under a V/III ratio as low as ~1.5, and then a thick AlN layer was grown at an increased temperature (1270 °C). Finally, an atomically flat AlN with a pit-free surface was obtained, confirmed by atomic force microscopy (AFM). In 2002, Shibata et al. [37] reported a single-step method for the growth of an AlN layer on sapphire, and the XRD rocking curve showed a very narrow full width at half-maximum (FWHM) of ~80 arcsec for the (002) direction. However, the FWHM of the XRD rocking curve for the (102) direction remained very large, typically ~1800 arcsec. The TEM measurement indicated that the dislocation density was around 10^{10} cm^{-2}, almost in edge type, which is different from GaN grown on sapphire. Recently, an AlN layer with an improved crystal quality was reported by the same group [38], and the FWHM of the XRD rocking curve for the (102) direction has been decreased to below 1000 arcsec. With the rapid development in the growth of high-crystal-quality AlN layer as the buffer on sapphire, it is no longer regarded as unrealistic to move the wavelength of LEDs down to UV-C.

Hanlon et al. [39] modified the two-step technique adopted by Ohba and Sato [36]. A thin AlN layer was initially deposited at a temperature as low as 700 °C under a low V/III ratio (~4.5), and then a thick AlN layer was grown at 1150 °C. This is similar to the currently standard two-step growth technique for GaN on sapphire. Based on this modified two-step technology, a FWHM of ~45 arcsec for the (002) XRD rocking curve was obtained, and a UV LED with an emission wavelength of 292 nm was then grown. In their structure, the thickness of the AlN buffer was ~0.7 μm, and a 6 nm single $Al_{0.4}Ga_{0.6}N$ quantum well was used as the emitting region, sandwiched by a 0.75 μm n-$Al_{0.69}Ga_{0.31}N$ bottom cladding layer and a 0.1 μm p-$Al_{0.64}Ga_{0.36}N$ top cladding layer. The turn-on voltage was ~8 eV, and the series resistance was ~50 Ω. The optical power was saturated to be 2 μW when the injection current was over 50 mA due to the heating effect. The first demonstration of a sub-300 nm UV LED was made by Adivarahan et al. in 2002 [40]. An AlN/AlGaN superlattice as a so-called dislocation filter was inserted into a 2 μm n-$Al_{0.4}Ga_{0.6}N$ buffer (not AlN) on sapphire. The active region consisted of a single 3 nm $Al_{0.32}Ga_{0.68}N$ quantum well, separated by 10 nm $Al_{0.36}Ga_{0.64}N$ barriers. A 20 nm p-$Al_{0.4}Ga_{0.6}N$ and a 50 nm partially relaxed p-GaN were finally grown. The partial strain relaxation of p-GaN was assumed to be beneficial for hole accumulation, leading to an increased hole concentration at the AlGaN/GaN interface. The optical power of the 285 nm emission was ~0.125 mW at 400 mA injection current, increasing to 0.25 mW at 650 mA after modifying the annealing conditions for both the p- and n-layer to obtain better p- and n-contact [41]. In addition, based on a similar structure, Fischer et al. [42] reported a 290 nm UV LED with an optical power of 0.12 mW at 20 mA, fabricated in a flipchip geometry. However, in all the above cases, there existed two other peaks at long wavelengths in addition to the main peak from the active region in the EL spectra. These long-wavelength peaks are associated with the p-AlGaN and p-GaN layers, which should be avoided by further optimization of device design and material growth.

Due to the high bond energy of Al–N, the mobility of Al adatoms is very low in the ambient of NH_3 at the currently used temperatures for growth of III–nitride semiconductor materials. In order to improve the crystal quality of AlN, it is necessary to increase the mobility of Al adatoms. A low flow rate of NH_3 has been proved to be one of the key parameters, as mentioned above. Currently, V/III is close to 1, meaning that it is difficult to make a further decrease. Further increasing the growth temperature seems to be very limited due to the current MOCVD technology. However, the further enhancement of the mobility of Al atoms can be realized by an approach evolved from the atomic layer epitaxial (ALE) technique, meaning that NH_3 and Al precursor (TMAl) are alternately supplied for a certain time period, allowing a beneficial surface migration of Al adatoms, which is very similar to the previous flow-rate modulation epitaxy also evolved from ALE for the growth of GaAs by Kobayashi et al. [43]. Khan et al. [44] were the first to apply this idea to the growth of GaN in 1992, and then extended it to the growth of AlInGaN quaternary [25] and AlN [45], leading to a significant improvement in the crystal quality of AlN. The FWHM of the (002) XRD rocking curve of AlN on sapphire was as narrow as 18 arcsec. Another technique to improve the crystal quality is to insert a short-period AlN/AlGaN superlattice mentioned above, originally aimed at controlling the strain [46]. Finally, it is found that the superlattice can also effectively decrease the dislocation density of the overlying AlGaN layer [47]. The reduction of dislocation density is due to dislocation bending, perhaps associated with small pits. A detailed TEM study indicated that a sub-superlattice has been observed in each AlGaN layer [48]. The injected supply of metal–organic precursors plays a role similar to silicon delta doping, which may introduce small pits. These pits may cause dislocation bending, and then annihilation, as previously observed in a silicon-doped Al-rich AlGaN layer.

With the application of both the above techniques, Sun et al. [49] reported a 280 nm UV LED with an output power of 0.85 mW at 20 mA injection current measured on a flipchip device, while the optical power measured on a bare chip was around 0.26 mW at 20 mA current, reported by Zhang et al. [50]. Ten short periods of $Al_{0.9}Ga_{0.1}N/Al_{0.7}Ga_{0.3}N$ superlattice were used to replace the previous $AlN/Al_{0.8}Ga_{0.2}N$ superlattice in order to achieve better strain relief, which also significantly improved surface morphology. Although the quantum-well thickness and Al composition remained the same as the previous one, five MQWs were used instead of the previous four MQWs as the active region. An increasing barrier of p-AlGaN electron blocking layer was believed to be helpful to suppress long-wavelength emission, thus a 25 nm p-$Al_{0.6}Ga_{0.4}N$ was used in order to improve the spectral purity. The I–V characteristic showed a turn-on voltage of ~6 V and a series resistance of ~14 Ω. The external quantum efficiency was improved to 0.93% in CW mode, and 1.28% in pulsed mode.

With modification of the top cladding layer, the electron blocking layer, and the MQWs as the active region, the same group [51] pushed the emission wavelength of the UV LEDs down to 265 nm. The active region consisted of five periods of silicon-doped 3 nm $Al_{0.45}Ga_{0.55}N$ quantum wells separated by 6 nm $Al_{0.55}Ga_{0.45}N$ barriers. The Al composition of the bottom cladding layer was increased to 60%, and the Al composition and thickness of the electron blocking layer were further increased to 65% and 50 nm, respectively. The CW optical power was 0.24 mW at 30 mA, and a maximum power of 10 mW was obtained at 1.2 A injection current in pulsed operation mode. The *I–V* curve indicated a turn-on voltage of ~6.8 V and a series resistance of 42 Ω, both of which were higher than for the 280 nm UV LEDs.

Unlike most other groups, Moe et al. [52] reported the successful growth of a 275 nm UV LED using an extremely high III/V ratio up to 23 336 on a SiC substrate. Also, the growth temperature was increased up to 1260 °C to enhance the migration mobility of the Al adatoms. Eventually, an optical power of 0.11 mW was obtained at 300 mA DC injection current, measured based on an interdigitated geometric device.

Recently, Allerman et al. [9] pushed the UV LEDs down to 237 nm but without reporting optical power. A standard structure was used, namely, an AlN buffer plus a short-period AlGaN/Al superlattice as a strain management/dislocation filter layer, and a thin (1–2 nm) $Al_{0.69}Ga_{0.31}N$ quantum well as the emitting region.

A 237 nm UV LED has been reported, but the performance of deep-UV LEDs is still poor; for example, the optical power of sub-250 nm LEDs is only a few µW [15]. In order to make further improvement, a lot of effort has to be put in to improving current spreading and light extraction in addition to crystal quality. With the push toward shorter wavelength, Al composition becomes higher, resulting in poorer conductivity due to an extremely large activation energy for the Mg acceptor; for example, a 320–390 meV activation energy for $Al_{0.70}Ga_{0.30}N$ [53]. The current spreading issue significantly affects performance. To overcome this issue, Wu et al. [54] developed a micropixel design for 254 nm UV LEDs in two classes: the first geometry consisting of a 4×4 array of pixels each with diameter 35 µm, and the second a 10×10 array of pixels each 25 µm in diameter. All of the LED micropixels were interconnected, allowing for a uniform injection current. This design led to a significant drop in the series resistance, from 80 Ω for a device in a standard size ($200 \times 200 \, \mu m^2$) to 17 Ω and 11 Ω for the 4×4 array of pixels and the 10×10 array of pixels, respectively. Fischer et al. [42] designed an interdigitated contact geometry in order to solve the issue of the saturation of optical power at high injection current, and obtained a CW output power of 1.34 mW for 290 nm UV LEDs at 300 mA. Jiang's group [122] reported circular UV LEDs, showing some improvements in solving current crowding and inhomogeneous driving. Generally, in semiconductor LEDs, about $1/(4n^2)$ of the light emitted from active region radiates through the top and bottom, where n is the index of refraction. In the case of LEDs containing DBRs, about $1/(2n^2)$ of the emitted light radiates through the top surface since the DBRs reflect the downward-propagating light. Therefore, the output power can be enhanced by a factor of ~2. However, with the emission wavelength of LEDs becoming shorter, it is more difficult to use DBRs since high refractivity requires a high refractive-index contrast, namely a high Al composition contrast of individual layers in one pair. The application of two-dimensional photonic crystals (PCs) in the fabrication of deep-UV LEDs is one of the best ways to enhance light extraction. The first application of PCs in III–nitride LEDs was reported by Jiang's group [55, 56]. Using PCs with a hole of diameter 200 nm and a lattice constant of 600 nm, the extraction for 330 nm UV LEDs was significantly improved by a factor of 2.5, as shown in Fig. 3.7, while the *I–V* curve remained almost the same as for the LEDs without using PCs.

However, the crystal quality – in particular, the crystal quality of the AlN buffer – remains a major issue. The ELOG technique has been proved to be an efficient way to yield a significant reduction of dislocation density in GaN films, but it may be difficult to apply in the growth of AlN films with low dislocation density due to the much larger sticking coefficient for Al than for Ga. AlN can easily nucleate and be grown on any mask, such as SiO_2 and Si_xN_y, currently used in the ELOG technology. Furthermore, the higher bond energy of Al–N (2.88 eV) in comparison to Ga–N (2.44 eV) [57] leads to a much lower lateral growth rate for AlN than for GaN because current commercial MOCVD machines have a limitation on the highest temperature available. Recently, Katona et al. [58] demonstrated a maskless ELOG growth technology. First, 0.6 or 1.0 µm of AlN was grown on sapphire, and then was dry etched using reactive ion etching to form stripes in the $\langle 1\bar{1}00 \rangle$ direction with a depth of ~0.4 µm as a mask for subsequent growth. Both mesa and trench were 5 µm wide. The subsequent growth of the overlying AlN layer was performed with a flow of a small amount of TMGa, which was expected to result in the enhancement of the lateral growth rate. Unfortunately, they were not successful in obtaining a fully coalesced overlying layer of $Al_{0.96}Ga_{0.04}N$, because the AlN nucleated even in the trench, and the trench material had the same vertical height as the wing materials due to a limited lateral growth rate. Both led to cessation of lateral growth. However, the TEM image indicated that there was a

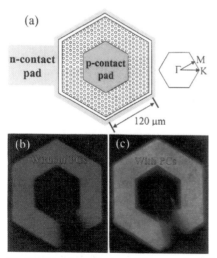

Fig. 3.7 (a) Schematic view of the mesa from the top, and (b) images of a LED without (left) and with PCs (right), both taken at 20 mA injection current. From Ref. [56].

significant reduction in dislocation density in the wing region. In early 2005, the same group [59] applied a similar technique to SiC substrate, and eventually a fully coalesced $Al_{0.93}Ga_{0.07}N$ was successfully obtained, as shown in Fig. 3.8. Compared to the case on sapphire substrate, there were a few improvements. First, the depth of the trench was significantly increased to 10 μm in order to avoid the cessation of lateral growth occurring in the case of the sapphire substrate, and the trench was made directly on the SiC substrate to form in the ⟨1$\bar{1}$00⟩ direction using an inductively coupled plasma system. Second, the flow rate of TMAl was decreased while that of TMGa was increased, both of which further improved the lateral growth rate to allow closure of the gap between the SiC ridges.

With further careful design of trench structure (width, depth, and gap) and further increase in growth temperature, the ELOG has been applied to the growth of AlN. Recently, Akasaki's group [60] reported the successful application of the ELOG in the growth of AlN on sapphire using a similar approach to that mentioned above but without the involvement of flowing TMGa. In order to improve the lateral growth rate, an extremely high temperature was used, like 1300 °C or even higher. A specially designed heater was employed, which might be difficult to achieve for current commercial MOCVD machines. Although a fully coalesced AlN was obtained, a new issue was raised, i.e. serious wafer warping at such high temperature. Therefore, a small size of sapphire substrate, not 2 inch sapphire, was used [61].

The above ELOG-related approaches are heavily involved in ex-situ patterning processes, and furthermore sapphire tends to be difficult to etch into a deep trench. More recently, Wang et al. [62] developed an approach for air-bridged lateral

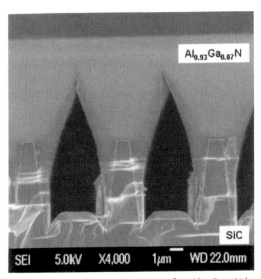

Fig. 3.8 Cross-sectional SEM image of an $Al_{0.93}Ga_{0.07}N$ layer film grown on an SiC substrate, demonstrating full coalescence. From Ref. [59].

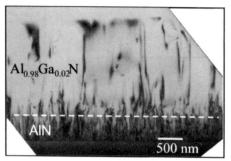

Fig. 3.9 Cross-sectional TEM image of a sample grown on a porous AlN layer, taken near the zone axis $\langle 1\bar{1}00 \rangle$ with $g = [11\bar{2}2]$. Clearly, the dislocation was stopped at the interface between the porous AlN and the overlying layer. From Ref. [62].

growth of an $Al_{0.98}Ga_{0.02}N$ layer by introduction of porosity in an AlN buffer, which did not involve any ex-situ patterning processes. First, a ~500 nm AlN layer with a high density of pores was directly grown on a sapphire substrate at 1150 °C. An AFM image of the porous AlN layer showed a high density of pores. The surface morphology of this porous AlN layer was critical for the subsequent growth of the overlying layer. Further decreasing the temperature could increase the area of the pores, making the subsequent growth of the overlying layer difficult to obtain an atomically flat surface. After that, the growth temperature was set to carry on growth of the AlN layer with a small amount of flowing Ga precursor (TMGa). The small TMGa flow was added in order to obtain enhancement of the lateral growth rate. The TEM image shown in Fig. 3.9, taken near the zone axis $\langle 1\bar{1}00 \rangle$ with $g = [11\bar{2}2]$, indicates that significant dislocation reduction has been made in the overlying layer. The XRD data in an (002) $\omega - 2\theta$ mode (Fig. 3.10) show that a ~98% Al composition in the overlayer has been formed due to a small amount of flowing TMGa. A small amount of Ga, such as 2%, added to an AlN layer, causes *only* ~2 nm redshift of bandgap emission, which is negligible in terms of fabrication of LEDs. The further evidence to demonstrate the highly improved crystal included an asymmetric XRD study, showing a significant decrease in the FWHM of the rocking curve. The same group also reported a highly improved optical property of MQWs grown on this highly improved layer, showing that the photoluminescence (PL) intensity of 265 nm at room temperature was significantly improved by one order of magnitude [63].

3.2.3
Ultraviolet Lasers

InGaN/GaN-based LDs with a lifetime of more than 10 000 h for CW operation at room temperature have been achieved. Recently, a blue–violet LD with an output power of over 1000 mW was also reported [64]. However, the emission wavelengths

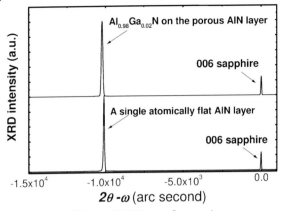

Fig. 3.10 The (002) $\omega - 2\theta$ XRD scan for samples grown on a porous AlN layer and a single atomically flat AlN layer for comparison. The (006) X-ray diffraction peak from sapphire is also included as a reference. From Ref. [62].

of these LDs were limited to between 390 and 465 nm. Even using low dislocation density GaN substrates, it is a challenge to fabricate UV LDs, in particular, when emission wavelength is shorter than 362 nm (i.e. GaN band-edge emission).

The first RT stimulated emission of III–nitrides in the UV spectral region was reported by Amano et al. [65] in 1990, which was realized under optical pumping using a pulsed nitrogen gas laser. The emission wavelength was around 375 nm. Their structure was very simple, only a single GaN layer on sapphire substrate grown using the standard two-step method. Although the first electrical injection III-nitride laser appeared in 1996, it took about five years to push the emission wavelength down to ~360 nm, reported by Nichia in 2001 [66]. The first ~360 nm UV LD was grown on a low dislocation density GaN substrate, achieved by the standard ELOG technique. Unlike InGaN-based LD, a single 10 nm GaN quantum well was used as an emitting region, sandwiched by 10 nm silicon-doped $Al_{0.15}Ga_{0.85}N$ barriers. Perhaps in order to avoid the formation of cracks, a low Al composition was used for both p- and n-cladding layers, consisting of 100 periods of $Al_{0.05}Ga_{0.95}N(2.5\,nm)/Al_{0.1}Ga_{0.9}N(2.5\,nm)$ superlattice. The Al composition for the guiding layers was also very low, 5% Al with a thickness of 150 nm in each for both top and bottom. As usual, a 10 nm p-$Al_{0.3}Ga_{0.7}N$ blocking layer was used to decrease carrier overflow. The device was fabricated in a standard ridge geometry. The cavity had 2.5 μm width and 600 μm length. The *I–V* and *I–L* characteristics showed a DC threshold current density of 3.5 kA cm^{-2} at a bias voltage of 4.6 V. The wavelength length was 366 nm in pulsed operation mode, and shifted to 369 nm in CW operation mode due to the heating effect. Shortly, they used an AlInGaN quaternary system as the emitting region, aiming to achieve better performance and shift toward shorter wavelength [67]. Instead of the 10 nm GaN quantum well, a single 10 nm AlInGaN quantum well was used as the emitting region. In order to improve the optical confinement, the Al compositions of individual layers in the superlattice as p- and n-cladding layer were increased to 9% and 13%, respectively. Also the numbers of periods for both p- and n-superlattices were increased equally to 125. Other structural parameters remained same. The *I–L* and *I–V* characteristics, measured based on a ridge geometry UV LD device, showed a DC threshold current density of 3.6 kA cm^{-2} at a bias voltage of 4.8 V, both of which were slightly higher than their first UV LD. The emission wavelength was at 365 nm.

Further moving toward shorter wavelength, Al compositions were increased for all the layers. Eventually, a 354 nm LD has been achieved in a pulsed operation mode; however, it suffered from a number of serious problems, including cracking and higher threshold current density of up to 14.3 kA cm^{-2}. A detailed study indicated that the crystal quality of the AlInGaN quaternary is a critical issue [68]. Figures 3.11 and 3.12 show the threshold current density as a function of Al and In composition in the AlInGaN active region, respectively, meaning that further increasing Al or In composition results in a higher threshold current density. Since the changes in both the optical confinement and carrier confinement are negligibly small for all investigated cases, the crystal quality of the AlInGaN should be mainly responsible for the significant change in threshold current density. This may be limited to using the AlInGaN system as the emitting region for shorter

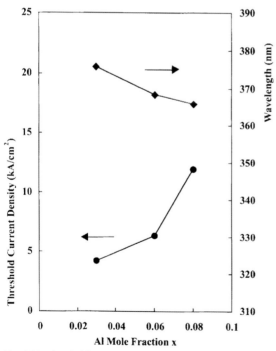

Fig. 3.11 Threshold current density and emission wavelength of AlInGaN SQW LDs as a function of Al mole fraction. From Ref. [68].

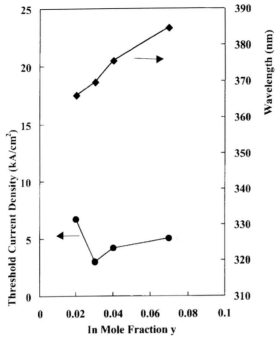

Fig. 3.12 Threshold current density and emission wavelength of AlInGaN SQW LDs as a function of In mole fraction. From Ref. [68].

wavelength. Therefore, in order to move to shorter wavelength, an AlGaN-based or AlGaN/GaN-based quantum well has to be employed. In 2003, Kneissl et al. [69] used three periods of 3.5 nm $Al_{0.02}Ga_{0.98}N$ MQWs as an active region to achieve a UV LD with an emission wavelength of 359.7 nm, working in a pulsed mode. In their structure, both n- and p-cladding layers were a short-period AlGaN superlattice with an average Al composition of 16%. The total thickness for the n-cladding layer was 0.6 µm, while that for the p-cladding layer was 0.5 µm. A 20 nm p-$Al_{0.3}Ga_{0.7}N$ blocking layer was usually used to decrease the carrier overflow. The threshold current density for their UV LD was ~23 kA cm^{-2}. Part of the high threshold current density was believed to be due to the crystal quality of the GaN buffer since their LD wafer was grown on sapphire without the involvement of any ELOG technique. Although the cracking issue was not mentioned, it was a great challenge since a higher Al composition was used in their structure. In order to avoid the formation of cracks and increase the crystal quality simultaneously, Akasaki's group [70] combined both the low-temperature AlN interlayer and the ELOG techniques, and finally obtained a thick and crack-free $Al_{0.18}Ga_{0.92}N$ layer with a low dislocation density on a sapphire substrate. In 2004, they were successful in fabricating a 350.9 nm UV LD using three periods of GaN(3 nm)/$Al_{0.08}Ga_{0.92}N$(8 nm): Si MQWs as the active region. Figure 3.13 shows their structure. The threshold current density was decreased to 7.3 kA cm^{-2}, much lower than the previous reports

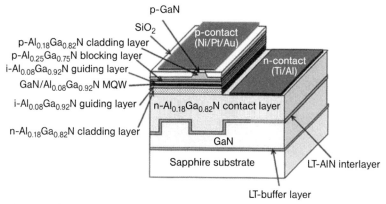

Fig. 3.13 Schematic structure of a UV LD grown on a low-dislocation-density AlGaN layer. From Ref. [70].

mentioned above. Recently, Cree further pushed the emission wavelength of UV LDs a little shorter, down to 343–348 nm [71], which is the shortest wavelength of the electrical injection LDs reported so far.

Obviously, there are a number of problems to be faced in realizing an electrical injection deep-UV LD, like cracking, doping, crystal quality, etc. The high crystal quality of AlN buffer instead of GaN buffer is very desirable, which can also solve the cracking issue to some degree. More importantly, it will ultimately determine lasing. So far, sub-330 nm UV lasing has been observed only under optical pumping. The first observation of the RT stimulated emission in the sub-330 nm spectral region was reported by Schmidt et al. in 1998 [72]. They used a dye laser to optically pump a single $Al_{0.26}Ga_{0.74}N$ layer grown on a sapphire substrate. Recently, similar to Khan's migration-enhanced MOCVD (MEMOCVD) method, Takano et al. [73, 74] used the flow-rate modulation approach, which has been modified for growth of AlN/GaN heterostructure and superlattice, reported by Hiroki and Kobayashi [75]. They significantly improved the crystal quality of AlN on SiC substrate. In order to obtain good strain control of AlN on SiC, a short-period AlN/GaN superlattice was initially grown as a buffer prior to AlN layer [76], this being different from the growth of AlN on sapphire where an AlN layer was first prepared prior to the growth of the superlattice as a strain-relief layer. The initial deposition of the short-period AlGaN/GaN superlattice also led to a significant reduction in the dislocation density of the AlN. Finally, UV lasing with an emission wavelength as short as 241.5 nm was achieved by optical pumping [73]. In their structure, a 0.6 μm $Al_{0.84}Ga_{0.16}N$ was used as a cladding layer, and a 0.1 μm $Al_{0.76}Ga_{0.24}N$ was used as a guiding layer. The active region consisted of three periods of 5 nm $Al_{0.76}Ga_{0.24}N$ quantum well separated by 10 nm $Al_{0.76}Ga_{0.24}N$ barriers. Actually, by modifying the Al composition in the MQWs, UV lasing has been demonstrated from 359.5 to 241.5 nm by optical pumping [74]. However, the threshold pumping power increased as a result of decreasing wavelength. Recently, the same group pushed the emission wavelength of UV lasing down to 228 nm by optical pumping [77], which is the shortest wavelength for lasing reported so far.

Compared to other III–V semiconductor LDs, the threshold current for III–nitride LDs is much higher. The high threshold current density is to a large degree an intrinsic limitation due to high carrier densities of states resulting from the heavy effective mass of the carriers in GaN-based materials. Generally, the threshold increases monotonically with increasing effective mass of the carrier. For example, the effective masses for GaN and GaAs are $0.19 m_0$ and $0.065 m_0$, leading to a threshold current density J_{th} of ~1 kA cm^{-2} for GaN-based QW LDs and 100 A cm^{-2} for GaAs-based QW LDs, respectively. In order to decrease the threshold, in theory, quantum dots (QDs) may have to be used as the emitting region. However, due to the heavy effective mass of the carriers in III–nitride materials, the size of QDs should be much smaller than that for other III–V semiconductor QDs. In that case, the population of carriers in the higher subband can be ignored, and thus the expected threshold current in both GaAs-based and GaN-based LDs should be almost the same at ~100 nA to 1 μA for a LD with a standard size [78, 79]. Therefore, with the use of QDs, the threshold current should be reduced by a factor of 100 in GaN-based LDs compared to a factor of 10 in GaAs-based LDs. Currently, despite a lot of reports on III–nitride QDs, the size of these dots may not be small enough, so that the predicted benefits of the use of dots, including low-threshold semiconductor lasers, have not yet been realized. Fabrication of a III–nitride polariton laser is another solution to massively decrease threshold current. A polariton is formed by electron–hole pairs (excitons) and photons and their interactions with one another [80]. Polaritons have a number of novel properties. The density of states in the light-emitting region is 10^4 times smaller than for excitons, thus offering the prospect of very low threshold current, at least an order of magnitude less than in conventional "photon" lasers. In order to realize the room-temperature polariton laser, the exciton must exist at room temperature. Therefore, a polariton laser can potentially be realized only using wide-bandgap materials, like III–nitride materials with a large exciton binding energy. As an initial first step toward the III–nitride polariton laser, high-reflectivity DBRs should be available, and second there should be a strong coupling between exciton and photon. So far, a high reflectivity up to 99.4% has been reported from violet to UV (360 nm) spectral region [31] and a strong coupling between exciton and photon at room temperature has also been observed in a III–nitride microcavity containing InGaN/GaN MQWs [81]. Although there are some other problems, the realization of a III–nitride polariton laser does not seem unrealistic any more.

3.3
InGaN-Based Emitters

The successful fabrication of ultra-bright InGaN-based violet/blue LEDs and long-lifetime InGaN-based LDs still remained major achievements in the III–nitride area over the last decade. Long-lifetime InGaN-based LDs have been used to fabricate "Blu-ray" DVDs, the next-generation DVD, which significantly increases the information storage capacity to 50 Gb in a dual-layer disk. Recently, NEC reported an over 1000 mW InGaN-based blue/violet LD in a planar inner stripe geometry [64] operating in a single mode. In the area of InGaN-based LEDs, Cree announced a 17 mW, 400 nm InGaN-based LED with a 28% external quantum efficiency in 2000 [82], and then shortly the output power was further increased to 21 mW. Combining a patterned substrate and flipchip technologies, in 2001 Tadatomo et al. [83] increased the optical power of a 382 nm InGaN-based LED up to 15.6 mW at 20 mA current, and to 38 mW at 50 mA, equivalent to a maximum external quantum efficiency of 24%. In 2004, Cree announced commercialized products of 460–470 nm LEDs with an optical power of 24 mW at 20 mA injection current, using a geometrically enhanced "epi-down" design [84]. Further improvements in violet/blue LEDs may be brought about by optimization of device design to allow the extraction of more light through improved device packages, such as photonic crystals reported by Jiang's group [55, 56], DBRs, etc.

However, when the emission shifts toward longer wavelength, in the green spectral region, the output power of InGaN LEDs is significantly decreased. For example, the output power was 8.5 mW for 505 nm LEDs, and decreased to 7 mW for 527 nm LEDs [84]. It is still a big challenge to make a further improvement of green LEDs from the point of view of material growth. Second, there is increasing interest in how to improve the technology for the growth of high-performance InGaN-based LEDs on silicon substrates, since this can benefit greatly from the current state-of-the-art technologies for silicon. The third issue is how to achieve white light without the involvement of any phosphor. Currently, there tends to be more efforts on these three areas in the field of InGaN-based LEDs.

In order to increase the indium composition for longer-wavelength devices, the growth temperature has to be reduced. Second, nitrogen has been proved to be necessary as carrier gas instead of H_2 for enhancement of indium incorporation. Both lead to a significant degradation in crystal quality, in particular, when a GaN layer is used as a potential barrier for an InGaN quantum well. Normally, the growth of a high-quality GaN layer requires a high temperature of more than 1000 °C and H_2 as carrier gas. An even worse issue is that the crack efficiency of NH_3 is greatly reduced at the low temperature required for the growth of the InGaN layer, which generally results in lack of nitrogen atoms even with a high V/III ratio. Thus the formation of indium droplets at temperatures below 800 °C is often observed [85]. Therefore, an extremely high V/III ratio is required to prevent the formation of indium droplets. For the growth of violet/blue LEDs, the indium composition is normally below 20%, and there may not be a major problem even though the GaN layer as barrier is grown under the optimized growth conditions for the InGaN layer. This is a general growth procedure for most of the current InGaN-based violet/blue LEDs. However, with wavelength shifting toward the green spectral region, this method, i.e. single-temperature growth of both GaN barrier and InGaN well, may cause a serious problem. In order to overcome this issue, a few groups reported the two-temperature technique to grow green LEDs [86–88]. Wen et al. [86] reported the temperature profile for growth of the active layer of their green LEDs. They first ramped the temperature down to 700 °C, waited until the temperature stabilized, and then grew the 3 nm thick InGaN well layer. Afterwards, they ramped the temperature up to 950 °C, waited until the temperature stabilized, and then grew the 10 nm GaN barrier layer. Five InGaN/ GaN MQWs as the active region were grown using the two-temperature method, and a 520 nm green LED with a maximum output power of 8.9 mW was obtained at 160 mA injection current. The same group further decreased the temperature to 680 °C for the growth of an InGaN well, and obtained a 579 nm LED at 20 mA injection current [87]. A detailed analysis was given by Thrush et al. [88], who compared the optical quality of InGaN/GaN MQWs over a wide spectral region from 400 to 580 nm grown using the single- and two-temperature methods, as shown in Fig. 3.14. The optical quality of these MQWs was judged by the FWHM of the emission peaks at 300 K. For the two-temperature method, the growth temperature was varied from 710 to 800 °C, and the temperature for growth of the GaN barrier was 900 °C. When the emission wavelength is shorter than 510 nm, the optical quality of the MQWs grown using the single-temperature method was superior to that using the two-temperature method. Above this, the situation became the opposite, meaning that the two-temperature method should be better than the single-temperature approach. Recently, based on the two-temperature method, a high efficiency of InGaN/GaN MQWs at 540 nm was reported by the same group [89].

Another obstacle to obtaining high output power of InGaN-based LEDs is the existence of a strong strain-induced piezoelectric field, which tilts the potential profile and then makes the electron and hole wavefunctions move in the opposite directions. As a result, the overlap between the electron and hole wavefunctions

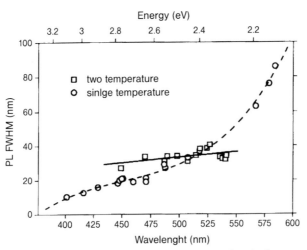

Fig. 3.14 PL FWHM as a function of emission wavelength of the InGaN/GaN MQW grown using the single- and two-temperature methods. From Ref. [88].

is reduced, resulting in a decrease in oscillator strength and recombination energy. This is known as the quantum-confined Stark effect, generally observed in InGaN/GaN MQWs. With increasing indium position to the green spectral region, the quantum-confined Stark effect becomes stronger under the assumption that quantum-well thickness remains the same. Typically, these devices demonstrate a blueshift in wavelength with increasing bias due to the screening effect of the strain-induced piezoelectric field. Chen et al. [87] reported a 23 nm blueshift of wavelength when the injection current was increased from 1 to 20 mA for a 2.5 nm $In_{0.7}Ga_{0.3}N$ well. Theoretically, the quantum-confined Stark effect sensitively becomes stronger as a result of increasing well thickness. Therefore, in order to fabricate high-performance InGaN-based LEDs, the active region should be carefully designed. Wang et al. [90] measured the output power of InGaN LEDs as a function of InGaN quantum-well thickness from 1.5 to 5 nm in order to study the influence of the quantum-confined Stark effect on optical performance, as shown in Fig. 3.15. In all cases, the structures were the same except for the thickness of the InGaN quantum well. In order to eliminate the possibility of strain relaxation, a single quantum well was chosen as the active region in all cases. The measurements were made under identical conditions. Figure 3.15 indicates that the output power decreases monotonically with increasing well thickness from 1.5 to 5 nm. Therefore, a thin InGaN quantum well may be better used as an active region, in particular for LEDs with long wavelength.

GaN grown on a silicon substrate suffers from a number of problems due to the large lattice mismatch (~20%) and large difference in thermal expansion coefficients between III–nitrides and silicon, such as the formation of cracks, high dislocation density, and cloudy surface morphology. All these result in a degraded optical performance including a low output power and a high bias voltage. Generally, (111) silicon is used as the substrate. A number of approaches have been

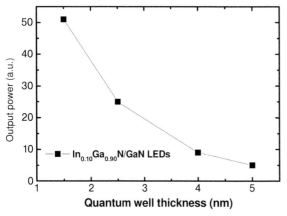

Fig. 3.15 Output power of the LEDs as a function of InGaN quantum-well thickness, measured under a 20 mA injection current at room temperature. From Ref. [90].

reported to avoid the formation of cracks and to increase the crystal quality. Dadgar et al. [91] reported a significant reduction in crack density by using a thin AlN interlayer grown at a low temperature, leading to growth of a crack-free GaN layer exceeding 1 μm in thickness. Feltin et al. [92] combined the AlN buffer and AlGaN/ GaN superlattice methods, expecting to obtain a better crystal quality. Finally, a 508 nm green LED was fabricated. The output power was ~6 μW at 20 mA injection current, and the applied bias voltage was very high, ~10.7 V at 20 mA current. Shortly after, using an in-situ SiN mask patterning technology, Dadgar et al. [93] reported that the crystal quality was greatly improved, confirmed by the (0002) XRD rocking curve and photoluminescence (PL) measurements. The TEM image indicated that the dislocation density was reduced to $10^9 \, cm^{-2}$ from the previous $10^{10} \, cm^{-2}$. This approach resulted in successful fabrication of a 455 nm blue LED with an optical power of 152 μW at 20 mA injection current. However, the optical power was saturated at 35 mA and then dropped, meaning a high series resistance, which implied that the crystal quality has to be improved. Ishikawa et al. [94] developed an intermediate buffer technology, i.e. an 80 nm thick AlN layer was initially deposited, and then a 0.25 μm thick $Al_{0.25}Ga_{0.75}N$ layer. Both were grown at a high temperature. A mirror-like surface was obtained, and the pits and cracks disappeared. The XRD data showed an improved crystal quality. Using this approach, the optical power of a 505 nm green LED was increased to 20 μW at 20 mA, and the applied bias voltage dropped to 7 V [95]. Optimization of a high-temperature AlN/GaN intermediate layer led to better crystal quality, resulting in a further drop of the applied bias voltage, for example, 4.1 V for green LEDs [96] and 3.7 V for blue LEDs [97], both at 20 mA. Recently, Shih et al. [98] reported the successful fabrication of both blue and green LEDs on Si substrates with output powers exceeding 0.7 mW at 20 mA with a turn-on voltage of 4–6 V. However, some cracks appeared. So far, the performance of all III–nitride devices is not comparable to those grown on sapphire or SiC substrates. The crystal quality of III–nitrides on Si remains a key issue, and how to manage the large strain is another great challenge.

The combination of III–nitride LEDs and appropriate phosphors remains the main technology to realize white light. However, the ideal method is to use a single chip to achieve white light in order to avoid using phosphors, as there are a number of issues, including limited lifetime and low conversion efficiency, in the application of phosphors. The use of a multiple-color LED in a single chip is obviously one good solution. Damilano et al. [99] used two different kinds of InGaN/ GaN MQWs as two active regions, and obtained a blue light at 470 nm, and another at 550 nm. Finally, they demonstrated a monolithic white LED at 20 mA current with a color temperature of 8000 K. In their structure, the blue emission was from 2.4 nm InGaN/GaN MQWs, and the 550 nm yellow emission was from 4.5 nm InGaN/GaN MQWs, both in the same indium composition. Generally, the quantum efficiency for the blue emission is much higher than that for the yellow emission. In order to well balance the EL intensities of these different color intensities, they used two periods of InGaN/GaN MQWs for the blue emission, and four periods of InGaN/GaN MQWs for the yellow emission. Nevertheless, the EL intensity of the blue emission was still much lower than that of the yellow emission. Yamada et al. [100] made a further development to realize white light by using a three-color LED. The active regions in their structure consisted of two periods of InGaN MQWs for red light, a single quantum well for green light, and two periods of InGaN MQW for blue light in order to make a best color match. At 20 mA, the color temperature, luminous flux, and luminous efficiency were improved to 5060 K, 0.532 lm, and 7.94 lm W^{-1}, respectively. Although monolithic white LEDs have been demonstrated, some key issues remain. In addition to the much lower luminous efficiency than that for the currently commercially available phosphor-based white LEDs, there exists a color rendering issue; for example, a blueshift in wavelength generally happens with increasing driving current, which becomes even worse for the green and red emissions. Therefore, such a three-color white LED can be used only under a particular driving current, which will be largely limited in the application. Therefore, an alternative approach should be developed from the point of view of material growth. More recently, researchers from Meijo University, Japan, used a purple-emitting LED combined with a "structured" SiC substrate that converts this emission into white light, and doubled the efficiency of current white LEDs to 130 lm W^{-1} [101]. Based on their report, it seemed that their approach did not involve a phosphor.

3.4
Nonpolar and Semipolar III–Nitride Emitters

So far, the major achievements in III–nitride optoelectronics have been made predominantly on (0001) sapphire or SiC substrates, resulting in these devices being grown along the polar (0001) direction. All the emitters discussed in Sections 3.2 and 3.3 were grown along the polar (0001) direction. Therefore, these devices suffered from strong polarization-induced internal electric fields, leading to a reduced overlap between the electron and hole wavefunctions and a low radiative recombination time as well as a blueshift in wavelength with increasing applied bias. It becomes even worse when the emission shifts toward longer wavelength, like either green or even red LEDs, since higher indium is required, as mentioned above. In order to overcome this obstacle, the growth of the III–nitride device along nonpolar directions was proposed. Takeuchi et al. [102] were the first to study the influence of crystal orientation on the piezoelectric field in GaN/InGaN quantum-well structures, and calculated the longitudinal piezoelectric field and transition probability in a strained 3 nm In$_{0.1}$Ga$_{0.9}$N/GaN quantum-well structure as a function of the polar angle from the (0001) direction, as shown in Figs. 3.16 and 3.17. Figure 3.18 shows a schematic illustration of the different crystal directions of GaN. The highest piezoelectric field is along the (0001) direction, resulting in the lowest transition probability, which happens in the current III–nitride devices. However, the piezoelectric field is reduced to almost zero, and the transition probability is significantly increased by a factor of 2.3, along the planes inclined at 39° from the (0001) direction, the highly symmetrical orientations,

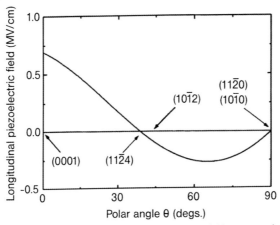

Fig. 3.16 Calculated longitudinal piezoelectric field in strained
$Ga_{0.9}In_{0.1}N$ on a GaN layer as a function of the polar angle
from the (0001) direction. From Ref. [102].

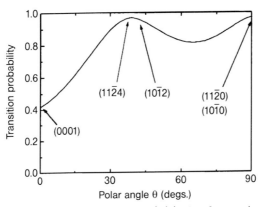

Fig. 3.17 Calculated transition probability in a 3 nm strained
$Ga_{0.9}In_{0.1}N$/GaN quantum well as a function of the polar angle
from the (0001) direction. From Ref. [102].

namely, (11$\bar{2}$4) and (10$\bar{1}$2). The other highly symmetrical orientations are (11$\bar{2}$0)
and (10$\bar{1}$0), along the planes inclined at 90° from the (0001) direction, which also
cause a zero piezoelectric field and an increase in transition probability by a factor
of 2.3. These four orientations are termed "nonpolar planes"; especially, the (11$\bar{2}$0)
and (10$\bar{1}$0) are labeled as *a*-plane and *m*-plane, respectively. In addition to benefit-
ing from zero piezoelectric field, devices grown along the nonpolar directions offer
a number of advantages over devices currently grown along the (0001) direction.
McLaurin et al. [103] reported a highly increased hole concentration up to 7 ×
10^{18} cm^{-2} in a p-type *m*-plane GaN layer, almost one order of magnitude higher

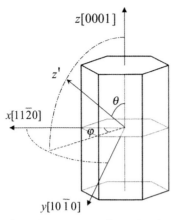

Fig. 3.18 Orientation relationships between (0001), (11$\bar{2}$0) and (10$\bar{1}$0) directions. The (11$\bar{2}$0) plane is called the *a*-plane, and the (10$\bar{1}$80) plane is the *m*-plane. From Ref. [102].

than for a standard p-GaN grown along the (0001) direction. Tsuchiya et al. [104] reported a similar result on a p-type *a*-plane GaN layer, and showed an activation energy of 118 meV for the Mg acceptor, much lower than the typical value of 150–170 meV for the p-GaN layer grown along the polar (0001) direction. This may be the reason for the highly increased hole concentration in the p-type *a*- or *m*-plane GaN. Park [105] predicted that the effective hole mass in a strained nonpolar InGaN quantum well should be smaller than that in a strained *c*-plane InGaN quantum well, which will potentially increase the hole mobility and thus improve the conductivity.

The first nonpolar GaN was grown on LiAlO$_2$ substrate via molecular-beam epitaxy (MBE) technology [106]. It may be difficult to grow GaN on LiAlO$_2$ substrate via MOCVD due to the high diffusivity of Li and Al atoms at the high temperature required for growth of III–nitrides. Generally, *a*-plane GaN can be grown using an *a*-plane GaN substrate obtained by hydride vapor-phase epitaxy (HVPE) or simply grown on an *r*-plane sapphire substrate. For the latter, the standard two-step technique, previously successful for the growth of a GaN layer on (0001) sapphire, can be applied. Figure 3.19 shows a schematic illustration of the epitaxial relationship between GaN and *r*-plane sapphire [117]. It has been shown that *m*-plane GaN can be grown on an *m*-plane GaN template obtained by HVPE. Recently, there have been several reports of LEDs prepared on a nonpolar *a*- or *m*-plane GaN. These devices showed almost zero blueshift of the emission wavelength with increasing driving currents, indicating the absence of polarization-induced electric fields, as expected. Chen et al. [107] reported a UV LED grown on *r*-plane sapphire. In their structure, three periods of GaN(6 nm):well/Al$_{0.12}$Ga$_{0.88}$N(12 nm): barrier MQWs were used as an active region. Although the EL spectrum showed an emission peak at 363 nm, the PL measurement of the MQW showed an emission peak

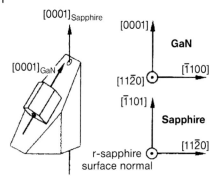

Fig. 3.19 Schematic illustration of the epitaxial relationship between GaN and *r*-plane sapphire. From Ref. [117].

at 360 nm, confirming the nonpolar effect, as the emission wavelength for such MQWs should be longer than 362 nm (i.e. GaN band-edge emission) in the case of GaN grown along the (0001) polar direction. Chitnis et al. [108] reported InGaN-based blue/purple LEDs grown on *r*-plane sapphire. They used three periods of $In_{0.15}Ga_{0.85}N(3.5\,nm)/GaN(10\,nm)$ MQWs as the active region. Although a blueshift was observed with increasing driving current from 2 to 50 mA, it was determined that the band-filling effect should be responsible for this blueshift. Their argument was based on the comparison of two individual *a*-plane LEDs but with very different well thicknesses, which showed a very similar blueshift in emission wavelength. In the case of such LEDs grown along the (0001) polar direction, a LED with a wide InGaN well should exhibit a much larger blueshift than that with a narrow InGaN well under the assumption of being fully strained for both. Shortly after, Chakraborty et al. [109] reported a nearly zero blueshift of the emission wavelength with increasing driving currents. The active region consisted of five periods of InGaN(4 nm):well/GaN(16 nm):barrier MQWs. The optical power at 413.5 nm was 240 μW at 20 mA injection current, equivalent to an external quantum efficiency of 0.4%. The InGaN-based LEDs grown along an *m*-plane GaN template showed a very similar result. Recently, Chakraborty et al. [110] reported a 450 nm InGaN-based LED grown on a free-standing *m*-plane GaN template, and obtained an optical power of 240 μW at 20 mA injection current. Recently, the issue of polarization anisotropy of nonpolar InGaN-based LEDs has been studied separately by a number of groups [111–113]. Generally, the polarization anisotropy is studied in terms of polarization ratio, labeled as ρ. The polarization ratio is defined as $\rho = |I_{\perp} - I_{\parallel}| / |I_{\perp} + I_{\parallel}|$, where $I_{\perp}$ and $I_{\parallel}$ are the EL or PL intensities with polarization in the direction perpendicular to the *c*-axis (i.e. π-polarization) and parallel to the *c*-axis, respectively. Sun et al. [111] reported a polarization ratio of 96%, based on *m*-plane InGaN/GaN MQWs grown on a γ-LiAlO$_2$ substrate by means of PL measurements. Gardner et al. [112] measured the polarization ratio of an *m*-plane InGaN-based LED grown on a (10$\bar{1}$0) plane 4H-SiC substrate. At a CW current density of 20 A cm^{-2}, the polarization ratio was 0.81, which decreased to 0.78 at

CW current density of 800 A cm^{-2}, while it increased to 0.9–0.88 in a pulsed measurement. The increased polarization ratio was attributed to the reduction of the heating effect in the pulsed measurement. More recently, Masui et al. [113] reported the polarization ratio of an *m*-plane InGaN-based LED grown on a freestanding *m*-plane GaN substrate, and obtained a low polarization ratio, only 0.17. They attributed the low polarization ratio to light scattering due to the thick p-metal layer, based on an argument that a higher polarization ratio of 0.42 was yielded when the measurement was made from the backside of the device, which could reduce light scattering.

An alternative approach to effectively reduce or possibly eliminate polarization effects is to grow III–nitride devices on semipolar planes, the orientation of which is along the planes with nonzero *h* or *k* Miller indices and nonzero *l* Miller indices. One advantage of growth along semipolar planes over nonpolar planes is that it offers more stability over a wide range of growth conditions [114]. Recently, Chakraborty et al. [115] reported a 439 nm InGaN-based LED grown on a semipolar (10$\bar{1}\bar{1}$) and (10$\bar{1}$3) GaN template, and obtained an output power of 190 μW at 20 mA. The EL spectra showed a driving-current-independent emission wavelength from 10 to 150 mA. More recently, Sharma et al. [116] reported a 525 nm green LED on a (10$\bar{1}$3) GaN template, and an output power of 19.3 μW at 20 mA was yielded, equivalent to an external quantum efficiency of 0.041%. The driving current induced blueshift in emission wavelength is only 6.7 nm over the 230 mA range, which was attributed to the band-filling effect.

Compared with the state-of-the-art technologies established for growth of III–nitride devices along the polar (0001) direction, the current growth technologies for devices along the nonpolar or semipolar directions are far from satisfactory. Crystal quality is still a key issue. Threading dislocation and stacking faults dominate the microstructure of the *a*-plane GaN grown on *r*-plane sapphire. The dislocation density is typically as high as (2–3) × 10^{10} cm^{-2} [117]. The ELOG and its evolved techniques remain the main approaches to reduce dislocation density. A detailed study by Craven et al. [118] indicated that the best orientation for deposition of the SiO$_2$ stripe was along the [$\bar{1}$100] direction for overgrowth of *a*-plane GaN, offering well-behaved and symmetric morphologies. More importantly, it does not lead to the dislocation bending into the wing regions, which is critical for a reduction of threading dislocation density. Chen et al. [119] used the ELOG technique twice to reduce the dislocation density of *a*-plane GaN. First, an SiO$_2$ stripe was deposited along the [$\bar{1}$100] direction, and then the wafer was overgrown to form an *a*-plane GaN stripe with vertical sidewall. The second overgrowth was performed after the deposition of SiO$_2$ both on top of the GaN stripes and in the gaps between the GaN stripes in order to suppress the vertical growth completely. The second overgrowth is very similar to the previously "pendeo-epitaxial" technique [120], also evolved from the ELOG. Eventually, the entire surface with full coalescence was composed of the lateral overgrown layer. An increased quantum efficiency has been observed on the AlGaN/GaN MQWs on an *a*-plane GaN template [121] prepared by the ELOG technique. In fact, most of the current structures including LEDs along the nonpolar or semipolar planes were grown on ELOG-prepared substrates. However, the optical performance is much worse, in comparison with devices grown along the polar (0001) direction, perhaps meaning that the crystal quality has to be further improved and that more efforts should be put in to investigation of the growth mechanism and emission mechanism.

3.5
Summary

Research activity on "lattice-mismatched heteroepitaxy" started a long time ago, with early GaP grown on silicon and GaAs on silicon, and now with III–nitrides on sapphire. There is no doubt that the development of the latter is much faster than the former. It is amazing to see such tremendous achievements in the growth of III–nitride optoelectronics within the last decade, in particular, in the "violet/ blue" technology, which has been commercialized. Further efforts have to be put in to the development of new technologies for fabrication of UV emitters and InGaN-based green emitters with high output power. In this chapter, we have reviewed the recent progress in UV emitters, high-power green emitters, UV-to-visible emitters grown on a silicon substrate or along the nonpolar or semipolar directions, and white LEDs using a single chip without the involvement of any phosphors. As one sort of "lattice-mismatched heteroepitaxy", the crystal quality of III–nitrides on sapphire still remains a key issue. Therefore, a high-quality GaN or AlN substrate with different orientations is necessary, which is expected to lead to further improvement in III–nitride devices with higher efficiency. An alternative precursor with a high crack efficiency to provide nitrogen atoms may significantly improve the growth of long-wavelength emitters. For growth facilities, the development of a more efficient heater to achieve a higher temperature may be beneficial to growth of an AlN layer with higher quality. Along with efforts on the development of technology, fundamental research on new physics has to be enhanced, which may result in a breakthrough in designing a new device based on the novel properties of III–nitride semiconductors.

Acknowledgments

The author acknowledges the support of the EPSRC (UK) through grant numbers EP/ C543521/1 and EP/C543513/1.

References

1 H. Amano, M. Kito, K. Hirmatsu, and I. Akasaki, *Jpn. J. Appl. Phys.* **1989**, 28, L2112.

2 S. Nakamura, T. Mukai, and M. Senoh, *Appl. Phys. Lett.* **1994**, 64, 1687.

3 E. D. Jones, *Light Emitting Diodes (LEDs) for General Illumination*, Optoelectronics Industry Development Association (OIDA), March, **2001**.

4 News, *Compound Semiconductor*, 3 March, **2004**.

5 News, *Compound Semiconductor*, 3 November, **2005**.

6 See, for example, "The global market for HB-LEDs", http://www.nanomarkets.net/

7 G. Rees, private communication.

8 X. Hu, J. Deng, J. P. Zhang, A. Lunev, Y. Bilenko, T. Katona, M. Shur, R. Gaska, M. Shatakov, and M. A. Khan, *6th Int. Conf. on Nitride Materials*, Bremen, Germany, 28 August–02 September, **2005**, Abstract Tu-PD2-2.

9 A. A. Allerman, M. H. Crawford, A. J. Fisher, K. H. A. Bogart, S. R. Lee, D. M. Follstaedt, P. P. Povencio, and D. D. Koleske, *J. Cryst. Growth* **2004**, 272, 227.

10 "GaN laser diodes: markets and applications", *Strategy Analytics*, March, **2004**.

11 T. Mukai and S. Nakamura, *Jpn. J. Appl. Phys.* **1999**, 38, 5735.

12 I. Akasaki and H. Amano, *J. Electrochem. Soc.* **1994**, 141, 2266.

13 Y. Kuga, T. Shirai, M. Haruyama, H. Kawanishi, and Y. Suematsu, *Jpn. J. Appl. Phys.* **1995**, 34, 4085.

14 J. Han, M. H. Crawford, R. J. Shul, J. J. Figiel, M. Banas, L. Zhang, Y. K. Song, H. Zhou, and A. V. Nurmikko, *Appl. Phys. Lett.* **1998**, 73, 1688.

15 J. P. Zhang, X. Hu, J. Ding, Y. Benlinko, A. Lunev, T. M. Katona, R. Gaska, and M. A. Khan, *Mater. Res. Soc. 2005 Fall Meeting*, Boston, 28 November–2 December, Symposium FF, Abstract, **2005**.

16 N. Otsuka, A. Tsujimura, Y. Hasegawa, G. Sugahara, M. Kume, and Y. Ban, *Jpn. J. Appl. Phys.* **2000**, 39, L445.

17 A. Kinoshita, H. Hirayama, M. Ainoya, Y. Aoyagi, and A. Hirata, *Appl. Phys. Lett.* **2000**, 77, 175.

18 T. Nishida, H. Saito, and N. Kobayashi, *Appl. Phys. Lett.* **2001**, 78, 399.

19 T. Nishida, H. Saito, and N. Kobayashi, *Appl. Phys. Lett.* **2001**, 79, 711.

20 M. Iwaya, S. Terao, T. Sano, S. Takanami, T. Ukai, R. Nakamura, S. Kamiyama, H. Amano, and I. Akasaki, *Phys. Stat. Sol. (a)* **2001**, 188, 117.

21 M. Iwaya, S. Takanami, A. Miyazaki, Y. Watanabe, S. Kamiyama, H. Amano, and I. Akasaki, *Jpn. J. Appl. Phys.* **2000**, 42, 400.

22 T. Wang, Y. Morishima, N. Naoi, and S. Sakai, *J. Cryst. Growth* **2000**, 213, 188.

23 Y. B. Lee, T. Wang, Y. H. Liu, J. P. Ao, Y. Izumi, Y. Lacroix, H. D. Li, J. Bai, Y. Naoi, and S. Sakai, *Jpn. J. Appl. Phys.* **2002**, 41, 4450.

24 Y. B. Lee, T. Wang, Y. H. Liu, J. P. Ao, H. D. Li, H. Sato, K. Nishino, Y. Naoi, and S. Sakai, *Jpn. J. Appl. Phys.* **2002**, 41, L1037.

25 J. P. Zhang, V. Adivarahan, H. M. Wang, Q. Fareed, E. Koukstis, A. Chitnis, M. Shatalov, J. W. Yang, G. Simin, M. A. Khan, M. S. Shur, and R. Gaska, *Jpn. J. Appl. Phys.* **2001**, 40, L921.

26 T. Wang, Y. H. Liu, Y. B. Lee, J. P. Ao, J. Bai, and S. Sakai, *Appl. Phys. Lett.* **2002**, 81, 2508.

27 H. Hirayama, K. Akita, T. Kyono, T. Nakamura, and K. Ishibashi, *Jpn. J. Appl. Phys.* **2004**, 43, L1241.

28 T. Wang, R. J. Lynch, P. J. Parbrook, R. Butté, A. Alyamani, D. Sanvitto, D. M. Whittaker, and M. S. Skolnick, *Appl. Phys. Lett.* **2004**, 85, 43.

29 A. Alyamani, D. Sanvitto, T. Wang, P. J. Parbrook, D. M. Whittaker, I. M. Ross, and M. S. Skolnick, *Phys. Stat. Sol. (c)* **2005**, 2, 813.

30 T. Wang, P. J. Parbrook, C. N. Harrison, J. P. Ao, and Y. Ohno, *J. Cryst. Growth* **2004**, 267, 583.

31 E. Feltin, J.-F. Carlin, G. Christmann, R. Butté, N. Grandjean, and M. Ilegems, *6th Int. Conf. on Nitride Materials*, Bremen, Germany, 28 August–2 September, **2005**, Abstract Th-OP4-1.

32 M. A. Khan, V. Adivarahan, J. P. Zhang, C. Q. Chen, E. Kuokstis, A. Chitnis, M. Shatalov, J. W. Yang, and G. Simin, *Jpn. J. Appl. Phys.* **2001**, 40, L1308.

33 A. Chitnis, V. Adivarahan, M. Shatalov, J. P. Zhang, M. Gaevski, W. Shuai, R. Pachipulusu, J. Sun, K. Simin, G. Simin, J. W. Yangand, and M. A. Khan, *Jpn. J. Appl. Phys.* **2002**, 41, L320.

34 H. Hirayama, *J. Appl. Phys.* **2005**, 97, 091101.

35 Y. Ohba and A Hatano, *Jpn. J. Appl. Phys.* **1996**, 35, L1013.

36 Y. Ohba and R. Sato, *J. Cryst. Growth* **2000**, 221, 258.

37 T. Shibata, K. Asai, T. Nagai, S. Sumiya, M. Tanaka, O. Oda, H. Miyake, and K. Hiramatsu, *Mater. Res. Soc. Symp. Proc.* **2002**, 693, I9.3.

38 T. Shibata, K. Asai, S. Sumiya, M. Mouri, M. Tanaka, O. Oda, H. Katsukawa, H. Miyake, and K. Hiramatsu, *Phys. Stat. Sol. (c)* **2003**, 0, 2023.

39 A. Hanlon, P. M. Pattison, J. F. Kaeding, R. Sharma, P. Fini, and S. Nakamura, *Jpn. J. Appl. Phys.* **2003**, 42, L628.

40 V. Adivarahan, J. P. Zhang, A. Chitnis, W. Shuai, J. Sun, R. Pachipulusu, M. Shatalov, and M. A. Khan, *Jpn. J. Appl. Phys.* **2002**, 41, L435.

41 V. Adivarahan, S. Wu, A. Chitnis, R. Pachipulusu, V. Mandavilli, M. Shatalov, J. P. Zhang, M. A. Khan, G. Tamulaitis, A. Sereika, I. Yilmaz, M. S. Shur, and R. Gaska, *Appl. Phys. Lett.* **2002**, 81, 3666.

42 A. J. Fischer, A. A. Allerman, M. H. Crawford, K. H. A. Bogart, S. R. Lee, R. J. Kaplar, W. W. Chow, S. R. Kurtz, K. W. Fullmer, and J. J. Figiel, *Appl. Phys. Lett.* **2004**, 84, 3394.

43 N. Kobayashi, T. Makimoto, Y. Yamaguchi, and Y. Horikoshi, *J. Appl. Phys.* **1989**, 66, 640.

44 M. A. Khan, J. N. Kuznia, R. A. Skogman, D. T. Olson, M. MacMillan, and W. J. Choyke, *Appl. Phys. Lett.* **1992**, 61, 2539.

45 J. P. Zhang, M. A. Khan, W. H. Sun, H. M. Wang, C. Q. Chen, Q. Fareed, E. Kuokstis, and J. W. Yang, *Appl. Phys. Lett.* **2002**, 81, 4392.

46 J. P. Zhang, H. M. Wang, M. E. Gaevski, C. Q. Chen, Q. Fareed, J. W. Yang, G. Simin, and M. A. Khan, *Appl. Phys. Lett.* **2002**, 80, 3542.

47 H. M. Wang, J. P. Zhang, C. Q. Chen, Q. Fareed, J. W. Yang, and M. A. Khan, *Appl. Phys. Lett.* **2002**, 81, 604.

48 W. H. Sun, J. P. Zhang, J. W. Yang, H. P. Maruska, M. A. Khan, R. Liu, and F. A. Ponce, *Appl. Phys. Lett.* **2005**, 87, 211915.

49 W. Sun, V. Adivarahan, M. Shatalov, Y. Lee, S. Wu, J. W. Yang, J. P. Zhang, and M. A. Khan, *Jpn. J. Appl. Phys.* **2004**, 43, L1419.

50 J. P. Zhang, X. Hu, Y. Bilenko, J. Deng, A. Lunev, R. Gaska, M. Shatalov, J. W. Yang, and M. A. Khan, *Appl. Phys. Lett.* **2004**, 85, 5532.

51 Y. Bilenko, A. Lunev, X. Hu, J. Deng, T. M. Katona, J. P. Zhang, R. Gaska, M. Shur, and M. A. Khan, *Jpn. J. Appl. Phys.* **2005**, 44, L98.

52 C. G. Moe, H. Masui, M. C. Schmidt, L. Shen, B. Moran, S. Newman, K. Vampola, T. Mates, S. Keller, J. S. Speck, S. P. DenBaars, C. Hussel, and D. Emerson, *Jpn. J. Appl. Phys.* **2005**, 44, L502.

53 M. L. Nakarmi, K. H. Kim, M. Khizar, Z. Y. Fan, J. Y. Lin, and H. X. Jiang, *Appl. Phys. Lett.* **2005**, 86, 092108.

54 S. Wu, V. Adivarahan, M. Shatalov, A. Chitnis, W. H. Sun, and M. A. Khan, *Jpn. J. Appl. Phys.* **2004**, 43, L1035.

55 T. N. Oder, K. H. Kim, J. Y. Lin, and H. X. Jiang, *Appl. Phys. Lett.* **2004**, 84, 466.

56 J. Shakya, K. H. Kim, J. Y. Lin, and H. X. Jiang, *Appl. Phys. Lett.* **2004**, 85, 142.

57 J. H. Edgar, *Properties of Group III Nitrides*, INSPEC, London, 1994, p. 74.

58 T. M. Katona, P. Cantu, S. Keller, Y. Wu, J. S. Speck, and S. P. DenBaars, *Appl. Phys. Lett.* **2004**, 84, 5025.

59 S. Heikman, S. Keller, S. Newman, Y. Wu, C. Moe, B. Moran, M. Schmidt, U. K. Mishra, J. S. Speck, and S. P. DenBaars, *Jpn. J. Appl. Phys.* **2005**, 44, L405.

60 (a) K. Nakano, M. Imura, G. Narita, T. Kitano, N. Fujimoto, N. Okada, K. Balakrishnan, M. Tsuda, M. Iwaya, S. Kamiyama, H. Amano, and I. Akasaki, *6th Int. Conf. on Nitride Materials*, Bremen, Germany, 28 August–2 September, **2005**, Abstract Th-P-028. (b) G. Narita, N. Fujimoto, Y. Hirose, T. Kitano, T. Kawashiwa, K. Iida, M. Tsuda, K. Balakrishnan, M. Iwaya, S. Kamiyama, H. Amano, and I. Akasaki, *ibid.*, **2005**, Abstract Mo-G1-3.

61 H. Amano, *2005 Annual Conf. of the British Association for Crystal Growth including Symposia on Semiconductors and Epitaxy Pharmaceuticals, Fine Chemicals and Polymers*, Sheffield, UK, 4–6 September, **2005**.

62 T. Wang, J. Bai, P. J. Parbrook, and A. G. Cullis, *Appl. Phys. Lett.* **2005**, 87, 151906.

63 T. Wang, J. Bai, P. J. Parbrook, and A. G. Cullis, *6th Int. Conf. on Nitride Materials*, Bremen, Germany, 28 August–2 September, **2005**, Abstract Fr-G8-2.

64 C. Sasaoka, K. Fukuda, M. Ohya, K. Shiba, M. Sumino, S. Kohmoto, K. Naniwae, M. Matsudate, E. Mizuki, I. Masumoto, R. Kobayashi, K. Kudo, T. Sasaki, and K. Nishi, *6th Int. Conf. on Nitride Materials*, Bremen, Germany, 28 August–2 September, **2005**, Abstract Th-PD3-1.

65 H. Amano, T. Asahi, and I. Akasaki, *Jpn. J. Appl. Phys.* **1990**, 29, L205.

66 S. Nagahama, T. Yanomoto, M. Sano, and T. Mukai, *Jpn. J. Appl. Phys.* **2001**, 40, L785.

67 S. Masui, Y. Matsuyama, T. Yanamoto, T. Kozaki, S. Nagahama, and T. Mukai, *Jpn. J. Appl. Phys.* **2003**, 42, L1318.

68 S. Nagahama, T. Yanomoto, M. Sano, and T. Mukai, *Jpn. J. Appl. Phys.* **2002**, 41, 5.

69 M. Kneissl, D. W. Treat, M. Teepe, N. Miyashita, and N. M. Johnson, *Appl. Phys. Lett.* **2003**, 82, 4441.

70 K. Iida, T. Kawashima, A. Miyazaki, H. Kasugai, S. Mishima, A. Honshio, Y. Miyake, M. Iwaya, S. Kamiyama, H. Amano, and I. Akasaki, *Jpn. J. Appl. Phys.* **2004**, 43, L499.

71 J. Edmond, A. Abare, M. Bergman, J. Bharathan, K. L. Bunker, D. Emerson, K. Haberern, J. Ibbetson, M. Leung, P. Russel, and D. Slater, *J. Cryst. Growth* **2004**, 272, 242.

72 T. J. Schmidt, Y. Cho, J. J. Song, and W. Yang, *Appl. Phys. Lett.* **1999**, 74, 245.

73 T. Takano, Y. Ohtaki, Y. Narita, and H. Kawanishi, *Jpn. J. Appl. Phys.* **2004**, 43, L1258.

74 T. Takano, Y. Narita, A. Horiuchi, and H. Kawanishi, *Appl. Phys. Lett.* **2004**, 84, 3567.

75 M. Hiroki and N. Kobayashi, *Jpn. J. Appl. Phys.* **2003**, 42, 2305.

76 Y. Ishihara, J. Yamamoto, M. Kurimoto, T. Takano, T. Honda, and H. Kawanishi, *Jpn. J. Appl. Phys.* **1999**, 38, L1296.

77 H. Kawanishi, M. Senuma, T. Nukui, Y. Ohtaki, and Y. Yamamoto, *6th Int. Conf. on Nitride Materials*, Bremen, Germany, 28 August–2 September, **2005**, Abstract Tu-PD-1.

78 Y. Arakawa, *Proc. SPIE* **2001**, 4580, 179.

79 Y. Arakawa, T. Someya, and K. Tachibana, *Phys. Stat. Sol. (b)* **2001**, 224, 1.

80 See, for example, A. Kavokin, G. Malpuech, and B. Gil, *MRS Internet J. Nitride Semicond. Res.* **2003**, 8 (3), Art. No. 3, and references therein.

81 T. Tawara, H. Gotoh, T. Akasaka, N. Kobayashi, and T. Saitoh, *Phys. Rev. Lett.* **2004**, 92, 256402.

82 News, *Compound Semiconductor*, **2000**, July 28.

83 K. Tadatomo, H. Okagawa, Y. Ohuchi, T. Tsunekawa, Y. Imada, M. Kato, and T. Taguchi, *Jpn. J. Appl. Phys.* **2001**, 40, L583.

84 http://www.cree.com

85 S. Keller, B. Keller, D. Kapolnek, U. Mishra, S. DenBaars, I. Shmagin, R. Kolbas, and S. Krishnankutty, *J. Cryst. Growth* **1997**, 170, 349.

86 T. C. Wen, S. J. Chang, Y. K. Su, L. W. Wu, C. H. Kuo, W. C. Lai, J. K. Sheu, and T. Y. Tsai, *J. Electron. Mater.* **2003**, 32, 419.

87 C. H. Chen, S. J. Chang, and T. K. Su, *Jpn. J. Appl. Phys.* **2003**, 42, 2281.

88 E. J. Thrush, M. J. Kappers, P. Dawson, M. E. Vickers, J. Barnard, D. Graham, G. Makaronidis, F. D. G. Rayment, L. Considine, and C. J. Humphreys, *J. Cryst. Growth* **2003**, 248, 518.

89 D. M. Graham, P. Dawson, M. J. Godfrey, M. J. Kappers, P. M. F. J. Costa, M. E. Vickers, R. Datta, and C. J. Humphreys, *6th Int. Conf. on Nitride Materials*, Bremen, Germany, 28 August–2 September, **2005**, Abstract Tu-P-053.

90 T. Wang, J. Bai, S. Sakai, and J. K. Ho, *Appl. Phys. Lett.* **2001**, 78, 2671.

91 A. Dadgar, J. Bläsing, A. Diez, A. Alam, M. Heuken, and A. Krost, *Jpn. J. Appl. Phys.* **2000**, 39, L1183.

92 E. Feltin, S. Dalmasso, P. de Mierry, B. Beaumont, H. Lahrèche, A. Bouillé, H. Haas, M. Leroux, and P. Gibart, *Jpn. J. Appl. Phys.* **2001**, 40, L738.

93 A. Dadgar, M. Poschenrieder, J. Bläsing, A. Diez, and A. Krost, *Appl. Phys. Lett.* **2002**, 80, 3670.

94 H. Ishikawa, G. Zhao, N. Nakada, T. Egawa, T. Jimbo, and M. Umeno, *Jpn. J. Appl. Phys.* **1999**, 38, L492.

95 T. Egawa, B. Zhang, N. Nishikawa, H. Ishikawa, T. Jimbo, and M. Umeno, *J. Appl. Phys.* **2002**, 91, 528.

96 B. Zhang, T. Egawa, H. Ishikawa, Y. Liu, and T. Jimbo, *Jpn. J. Appl. Phys.* **2003**, 42, L226.

97 T. Egawa, T. Moku, H. Ishikawa, K. Ohtsuka, and T. Jimbo, *Jpn. J. Appl. Phys.* **2002**, 41, L663.

98 C. F. Shih, N. C. Chen, C. A. Chang, and K. S. Liu, *Jpn. J. Appl. Phys.* **2005**, 44, L140.

99 B. Damilano, N. Grandjean, C. Pernot, and J. Massies, *Jpn. J. Appl. Phys.* **2001**, 40, L918.

100 M. Yamada, Y. Narukawa, and T. Mukai, *Jpn. J. Appl. Phys.* **2002**, 41, L246.

101 News, *Compound Semiconductor*, **2005**, November 22.

102 T. Takeuchi, H. Amano, and I. Akasaki, *Jpn. J. Appl. Phys.* **2000**, 39, 413.

103 M. McLaurin, T. E. Mates, and J. S. Speck, *Appl. Phys. Lett.* **2005**, 86, 262104.

104 Y. Tsuchiya, Y. Okadome, A. Honshio, Y. Miyake, T. Kawashima, M. Iwaya, S. Kamiyama, H. Amano, and I. Akasaki, *Jpn. J. Appl. Phys.* **2005**, 44, L1516.

105 S. H. Park, *J. Appl. Phys.* **2002**, 91, 9904.

106 P. Waltereit, O. Brandt, A. Trampert, H. T. Grahn, J. Menniger, M. Ramsteiner, M. Reiche, and K. H. Ploog, *Nature (London)* **2000**, 406, 865.

107 C. Chen, V. Adivarahan, J. W. Yang, M. Shatlov, E. Kuokstis, and M. A. Khan, *Jpn. J. Appl. Phys.* **2003**, 42, L1039.

108 A Chitnis, C. Chen, V. Adivarahan, M. Shatalov, E. Kuokstis, V. Mandavilli, J. W. Yang, and M. A. Khan, *Appl. Phys. Lett.* **2004**, 84, 3663.

109 A. Chakraborty, B. A. Haskell, S. Keller, J. S. Speck, S. P. Denbaars, S. Nakamura, and U. K. Mishra, *Appl. Phys. Lett.* **2004**, 85, 5143.

110 A. Chakraborty, T. J. Baker, B. A. Haskell, F. Wu, J. S. Speck, S. P. Denbaars, S. Nakamura, and U. K. Mishra, *Jpn. J. Appl. Phys.* **2005**, 44, L173.

111 Y. Sun, O. Brandt, M. Ramsteiner, H. T. Grahn, and K. H. Ploog, *Appl. Phys. Lett.* **2003**, 82, 3850.

112 N. F. Gardner, J. C. Kim, J. J. Wierer, Y. C. Shen, and M. R. Krames, *Appl. Phys. Lett.* **2005**, 86, 111101.

113 H. Masui, A. Chakraborty, B. A. Haskell, U. K. Mishra, J. S. Speck, S. Nakamura, and S. P. DenBaars, *Jpn. J. Appl. Phys.* **2005**, 44, L1329.

114 K. Nishizuka, M. Funato, Y. Kawakami, S. Fujita, Y. Narukawa, and T. Mukai, *Appl. Phys. Lett.* **2004**, 85, 3122.

115 A. Chakraborty, T. J. Baker, B. A. Haskell, F. Wu, J. S. Speck, S. P. Denbaars, S. Nakamura, and U. K. Mishra, *Jpn. J. Appl. Phys.* **2005**, 44, L945.

116 R. Sharma, P. M. Pattison, H. Masui, R. M. Farrell, T. J. Baker, B. A. Haskell, F. Wu, S. P. DenBaars, J. S. Speck, and S. Nakamura, *Appl. Phys. Lett.* **2005**, 87, 231110.

117 M. D. Craven, S. H. Lim, F. Wu, J. S. Speck, and S. P. DenBaars, *Appl. Phys. Lett.* **2002**, 81, 469.

118 M. D. Craven, S. H. Lim, F. Wu, J. S. Speck, and S. P. DenBaars, *Appl. Phys. Lett.* **2002**, 81, 1201.

119 C. Chen, J. P. Zhang, J. W. Yang, V. Adivarahan, S. Rai, S. Wu, H. Wang, W. H. Sun, M. Su, Z. Gong, E. Kuokstis, M. Gaevski, and M. A. Khan, *Jpn. J. Appl. Phys.* **2003**, 42, L818.

120 T. S. Zheleva, O. H. Nam, M. D. Bremser, and R. F. Davis, *Appl. Phys. Lett.* **1997**, 71, 2472.

121 T. Koida, S. F. Chichibu, T. Sota, M. D. Craven, B. A. Haskell, J. S. Speck, S. P. DenBaars, and S. Nakamura, *Appl. Phys. Lett.* **2004**, 84, 3768.

122 K. H. Kim, Z. Y. Fan, M. Khizar, M. L. Nakarmi, J. Y. Lin, and H. X. Jiang, *Appl. Phys. Lett.* **2004**, 85, 4777.

Part II

4
ZnSeTe Rediscovered: From Isoelectronic Centers to Quantum Dots

Yi Gu, Igor L Kuskovsky, and G. F. Neumark

4.1
Introduction

Among wide-bandgap semiconductors, ZnSe and related alloys occupy a histori-cally important niche since they were the first to exhibit lasing in the green–blue spectral region (see e.g. Ref. [1] and references therein). As good bipolar doping is required for light-emitting device (LED) applications, Zn–Se–Te systems were recognized early due to the fact that ZnSe can be easily doped n-type and ZnTe is easily doped p-type, and this has led to extensive studies on ZnSeTe alloys for more than 25 years (see e.g. Akimova et al. [2]).

The latest advances in the doping of Zn–Se–Te have been discussed recently [3, 4] and the technique leading to such advances has been applied to obtaining undoped Zn–Se–Te systems showing very interesting optical properties. Here, we shall address the new developments in these systems, with the focus on a novel Zn–Se–Te multilayers grown by the migration-enhanced epitaxy (MEE) technique, where Te is introduced in submonolayer quantities. This leads to the simultaneous formation of both Te isoelectronic centers and quantum dots (QDs), which gives rise to many interesting properties as discussed in this chapter.

In general, isoelectronic centers are impurity states that are formed when the substitutional atoms have the same valence as the host atoms they replace but where the substitutional atoms have a different electronegativity and/or size. These isoelectronic centers, which can take various forms of single atoms, non-nearest-neighbor atom clusters, or nearest-neighbor atom clusters, can introduce energy levels inside the bandgap. Upon excitation, e.g. by optical illumination, these localized levels can trap excitons, which recombine and emit luminescence. In bulk $ZnSe_{1-x}Te_x$ alloys, the substitutional Te atoms form such isoelectronic centers, and the photoluminescence (PL) from these materials is dominated by excitons bound to these isoelectronic centers [so-called isoelectronic bound exci-tons (IBE)]. The microscopic configuration of these isoelectronic centers has been

Wide Bandgap Light Emitting Materials and Devices. Edited by G. F. Neumark, I. L. Kuskovsky, and H. Jiang
Copyright © 2007 WILEY-VCH Verlag GmbH & Co. KGaA, Weinheim
ISBN: 978-3-527-40331-8

studied fairly extensively but still remains open to debate; however, it is commonly accepted that at least some of them are formed by two or more nearest-neighbor Te atoms (henceforth referred to as Te_n centers).

Recently there has been a renewed interest in impurity states in solids as they have been discovered to possess the properties of two-level quantum systems [5–7], that is, these impurity states emit single photons at predetermined times and with well-established spectral and temporal characteristics. This makes them appealing for quantum computation applications. Moreover, compared to other two-level quantum systems, including quantum dots and molecules, impurity states in solids have the advantages of avoiding the material issues usually associated with QDs (e.g. surface/interface defects) and also of easy circuit integration compared to molecules. Among impurity states, isoelectronic centers appear to be especially interesting as they are very similar to neutral QDs and (in the absence of free carriers) do not permit Auger exciton recombination. Indeed, in a very recent study, within a single ZnSe:Te monolayer, Muller et al. [8] have shown that sharp PL emissions (due to IBEs) first reported by Kuskovsky et al. [9] exhibit properties of a single-photon emitter. These promising results warrant further investigations into the microscopic origins of Te isoelectronic centers.

The next interesting and fundamental question in a system where there is a size distribution in isoelectronic centers, such as in Zn–Se–Te, is whether the large isoelectronic centers behave as QDs. Theoretical studies [10] have shown that large Indium isoelectronic centers in $In_xGa_{1-x}N$ systems possess the same electronic properties as QDs, and these studies also predicted a smooth transition in electronic properties between isoelectronic centers and QDs. Experimental studies on Zn–Se–Te, however, have almost exclusively attributed the PL emission to Te-related IBEs even in ZnTe/ZnSe multiple-quantum-well (MQW) structures [11] where Te exists in large quantities. Only very recently, the first direct evidence of QD-related PL has been presented by Gu et al. [12] for the Zn–Se–Te multilayer system, where the often observed PL peak at ~2.5 eV is shown to be due to multicenter emissions, among which are $Te_{n>2}$ isoelectronic centers and ZnTe/ZnSe QDs. These QDs are of a type-II nature, as expected [13]. Using PL excitation spectroscopy (see Fig. 4.14 and also below), Gu et al. [12] showed that there is a continuing change in the emission properties from isoelectronic centers to QDs, confirming the theoretical results [10]. It must be noted that type-II QDs possess many interesting optical properties, including theoretically predicted Aharonov–Bohm (AB) effect for a neutral particle [14–17]; the multilayered structure of ZnTe/ZnSe QDs discussed in this chapter has enabled the first observation of one manifestation of the AB phase experimentally [18].

The formation of QDs in this novel Zn–Se–Te multilayer system also raises interesting materials science issues. For instance, the growth mechanism of the QDs is of interest. On the one hand, due to the large lattice mismatch (~7%) between ZnTe and ZnSe, one would expect the Strakskii–Krastanow (SK) growth mechanism (two-dimensional to three-dimensional). However, the Volmer–Weber growth mechanism (three-dimensional) has also been suggested [19]. Alternatively, we suggest that, in our case, since Te is introduced only in submonolayer quantities, the SK growth mechanism also seems unlikely and QDs probably form via Te surface diffusion promoted by MEE. Further studies are needed to clarify the growth mode of QDs with only submonolayer depositions.

This chapter is organized as follows. After a brief description of growth procedures and sample structures, we first show high-resolution X-ray diffraction studies on the sample structural properties. Next we give a brief review of the optical proprieties of bulk dilute $ZnSe_{1-x}Te_x$ alloys. We then discuss the optical properties of the Zn–Se–Te multilayer system, including detailed studies on ZnTe/ZnSe QDs. Finally we present very recent results of magneto-optical studies, including the observation of the optical Aharonov–Bohm effect.

4.2
Sample Growth

This multilayer material system was grown by molecular-beam epitaxy (MBE) on (001) GaAs substrates in a Riber 2300 system, which includes III–V and II–VI growth chambers connected by ultrahigh vacuum (UHV). Oxide desorption of the GaAs substrates was performed in the III–V chamber by heating the substrates to 590 °C with an As flux impinging on the surface. Then, a 200 nm GaAs buffer layer was grown at 580 °C, and this gave a streaky (2 × 4) reflection high-energy electron diffraction (RHEED) surface pattern. The substrate with the GaAs buffer layer was transferred to the II–VI chamber under UHV. Prior to the growth of the II–VI epilayers, Zn irradiation of the GaAs surface was performed. This step is intended to suppress the formation of Ga_2Se_3 at the III–V/II–VI interface, which is believed to be related to the formation of stacking faults [20]. Then a 400 nm thick undoped ZnSe buffer layer was grown at 250 °C under Se-rich conditions with a growth rate of $0.8\,\mu m\,h^{-1}$. RHEED gives a streaky (2 × 1) pattern. After buffer layer growth, the multilayer structure was then grown with the Se, Zn, and Te shutter sequences shown in Fig. 4.1(a). Multiple monolayers (MLs) of a ZnSe spacer (>10 MLs) were grown by opening the Zn and Se shutters together, after which the Se shutter was closed for 5 s to produce a Zn-terminated surface. Then all shutters were closed for 5 s to desorb excess Zn from the surface. After this the Te shutter was opened for 5 s to deposit Te onto the Zn-terminated surface, and then all shutters were closed for 5 s followed by opening the Zn shutter for 5 s to produce another Zn-terminated surface. Two types of samples were grown, one containing one Zn–Te layer (denoted as δ-ZnSe:Te) and one containing three consecutive Zn–Te layers (denoted as δ^3-ZnSe:Te). After the deposition of Zn–Te layers, the Se shutter was then opened to start the next growth sequence. This sequence was repeated for several periods in order to obtain layers thick enough for measurement. The resulting schematic sample structures are plotted in Figs. 4.1(b) and (c) for δ-ZnSe:Te and δ^3-ZnSe:Te, respectively. For each of these types of sample structure, several samples were grown (see Table 4.1).

It should be noted that Te (Zn) is deposited without Zn (Te), a growth procedure known as migration-enhanced epitaxy (MEE), which enhances surface diffusion.

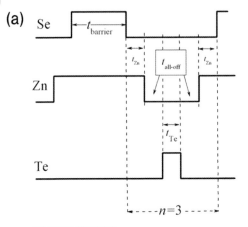

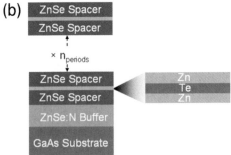

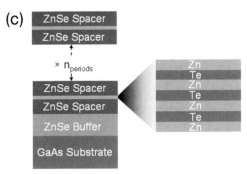

Fig. 4.1 (a) Shutter sequence for growth of Zn–Se–Te multilayer structures. (b, c) Schematics of Zn–Se–Te multilayer structures.

Since a very small Te flux is used during the deposition, only a fraction of a Te monolayer (submonolayer) is formed (see Section 4.3), and this leads to many interesting properties, with the optical properties discussed in Section 4.4 being of primary interest here.

Table 4.1 Sample structures.

Sample	Structure
A1	δ-ZnSe:Te
A2	δ-ZnSe:Te
A3	δ-ZnSe:Te
B1	δ^3-ZnSe:Te with flux ratio Te/Zn ~ 0.257
B2	δ^3-ZnSe:Te with flux ratio Te/Zn ~ 0.440
B3	δ^3-ZnSe:Te with flux ratio Te/Zn ~ 0.260

4.3
Structural Properties

The structural properties of this multilayer system have been investigated by high-resolution X-ray diffraction (HRXRD) measurements [21]. HRXRD is a very effective technique to study the structure of multilayer systems (including a multilayer heterostructure containing quantum dots) due to its high sensitivity to spatial long-range periodicity and strain distribution, as well as to its nondestructive characteristics [22–25]. With a large beam size (of the order of square millimeters), such X-ray diffraction experiments investigate the overall properties of epitaxial films. By analyzing HRXRD results [22, 25], the following parameters can be extracted by simulating X-ray diffraction curves: (i) periodicity; (ii) period dispersion [we here assume a Gaussian distribution of period T with standard deviation $\sigma(T)$]; (iii) individual layer thickness; and (iv) alloy compositions. Of particular interest here are the δ-layer thickness and Te compositions, which are related to optical properties.

X-ray measurements [21] were carried out at room temperature at Beamline X20A at the National Synchrotron Light Source (NSLS) at Brookhaven National Laboratory. Monochromatic synchrotron radiation at 8 keV ($\lambda = 1.54056$ Å), with a double-crystal Ge (111) monochromator and a Si (111) analyzer, was used to obtain symmetric (004) reflections in the $\omega - 2\theta$ mode on typical δ-ZnSe:Te [sample A2, in Fig. 4.2(a)] and δ^3-ZnSe:Te [sample B2, in Fig. 4.2(b)] samples [26]. The X-ray diffraction curves of both samples show a sharp peak located at about 66.006°, which is from the GaAs substrate. On both sides of the GaAs peak, superlattice (SL) satellites up to first (second) orders are visible for sample A2 (B2), and all observed SL satellites are distributed with nearly equal angular distances. Thus, the overall periodicity along the growth direction is preserved by introducing the Zn–Te δ layers. Compared with the zeroth-order satellite [SL(0)], higher-order satellites are broadened (Fig. 4.2), which indicates a period dispersion (i.e. fluctuation of periodicity) in our samples [22, 25].

In order to extract the structural parameters of these samples, the measured $\omega - 2\theta$ scans were be simulated [27, 28]. Table 4.2 lists all the structural parameters involved in the simulation process.

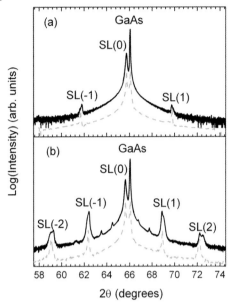

Fig. 4.2 X-ray diffraction (004) $\omega - 2\theta$ scan curves (solid lines) of (a) sample A2 and (b) sample B2, plotted in a semilogarithmic scale. The dashed lines are simulation results.

Table 4.2 Structural parameters involved in the XRD simulation process.

T	Period of multilayers
t_1	Thickness of the spacers
t_2	Thickness of Zn–Te layers
x_1	Te composition in the spacers
x_2	Te composition in Zn–Te layers
$\sigma(T)$	Period distribution, defined as standard deviation of period distribution

The simulation of the (004) $\omega - 2\theta$ curves was carried out using the dynamical model, which is preferred in this case since the GaAs single-crystal substrate is very thick and the overall thickness of the multilayer structure is at least ~0.6 µm, close to the extinction depth. The diffraction geometry is shown in Fig. 4.3. The simulated diffraction intensity is obtained as $|X(0)|^2$, where $X_j \equiv X(Z_j)$ is the diffraction coefficient [29] at a given depth below the surface of the film. The coefficient X_j can be calculated using the recurrence relation

$$X_j = \frac{wX_{j+1} - i(2C\chi_h + uX_{j+1})\tan[\Phi(z_{j+1} - z_j)/2]}{w + i(2gC\chi_{\bar{h}}X_j + u)\tan[\Phi(z_{j+1} - z_j)/2]} \qquad (4.1)$$

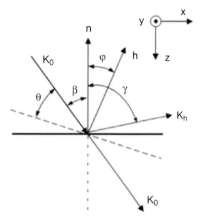

Fig. 4.3 The X-ray diffraction geometry.

which is the solution of the one-dimensional differential equation given by Halliwell et al. [30]. Here, C is the polarization factor, which is equal to 1 for σ-polarization and to $\cos 2\theta$ for π-polarization; χ_0, χ_h and $\chi_{\bar{h}}$ are the Fourier coefficients of the dielectric susceptibility [31]. The other parameters are defined as $\Phi = \pi K w / \cos \beta_B$, $g = \cos \gamma_B / \cos \beta_B$, $u = (1 + g)\chi_0 + 2\eta \sin \theta_B$, and $w = [u^2 - 4C^2 g \chi_h \chi_{\bar{h}}]^{1/2}$, where K is the wavenumber of the incident X-ray wave in vacuum, η is the deviation angle from the exact Bragg angle, β and γ are the angles of the incident beam and the diffracted beam in the normal direction to the surface **n** (see Fig. 4.3), respectively, and the subscript "B" means values at the exact Bragg condition.

For our specific system, we assume that the substrate is a perfect single crystal with infinite thickness; therefore, the diffraction coefficient at the interface of the substrate and the epitaxial film can be simply described by

$$X(z = L) = \frac{-u - w}{2gC\chi_{\bar{h}}} \quad \mathrm{Im}(w) > 0 \tag{4.2}$$

where L is the total thickness of the thin film. Since $X(L)$ represents the reflection from the substrate, to calculate $X(L)$ the GaAs structural parameters should be used. The diffraction coefficient at the surface, $X(0)$, can be calculated using the recurrence relation shown in Eq. (4.1), if the structural parameters for each sublayer in the epitaxial film are known. Then, the simulated diffraction intensity curve is generated as $|X(0)|^2$.

According to the simulation, the intensities of the high-order SL satellites depend on the Te concentration in the δ layers, i.e. the value of x_2. It is to be noted that, if x_2 is close to x_1, the high-order SL satellites will become very weak, which is apparently the case for the δ-ZnSe:Te sample [Fig. 4.2(a)]. This is expected because when $x_2 = x_1$, the difference between the spacer and the δ layers disappears, and no superlattice structure exists in the film. The measured diffraction curves for the δ³-ZnSe:Te [Fig. 4.2(b)] exhibit quite strong high-order SL peaks,

Table 4.3 Structural parameters obtained from the simulation.

Sample	t_1 (Å)	t_2 (Å)	x_1 (%)	x_2 (%)
A2	25.3	0.7	0.5	25
A3	24.1	0.7	1.5	30
B1	30.1	0.7	0.5	50 ± 15
B2	31.2	0.8	2	70 ± 15
B3	24.3	0.7	2	50 ± 15

which indicates that the δ layers have a much higher Te composition than that in the spacers, i.e. the Te is mainly "locked" in the δ layers. A difference in the diffraction curves from δ-ZnSe:Te [Fig. 4.2(a)] and δ³-ZnSe:Te [Fig. 4.2(b)] is expected considering that less Te is introduced into δ-ZnSe:Te.

As mentioned earlier, the higher-order satellites are broadened compared with the SL(0) satellite. It has been suggested [25] that a period variation along the growth direction can contribute to this broadening. Thus, we introduced a Gaussian period dispersion in the simulation with a standard deviation of ~3%. The best-fitting curves are shown in Fig. 4.2 as the dashed lines. The corresponding layer thicknesses as well as Te concentrations for several A and B series samples (see Table 4.1) are listed in Table 4.3.

There are several significant findings. (1) The δ layer can be seen to contain submonolayer quantities of Te for both series of samples, as expected from the growth conditions. (2) Although the growth conditions for the δ layers for these samples are different, the Te composition in the spacer region is low in all cases, in fact being approximately the same, and is presumably due to the Te background presence even with the Te shutter closed, suggesting that Te is "locked" in the δ layers. (3) Higher Te/Zn flux ratios (sample B2) result in a higher Te composition value in the δ layers compared to those of B1 and B3, the effect of which will be discussed in Section 4.4.

Combining the facts that the Te concentrations are high in the δ layer region and that the Te coverage is low (submonolayer), one can conclude that Te *does not* distribute uniformly in the plane of the δ layers, and it is thus very likely that there is a formation of Te-rich quantum dots. This gives rise to many interesting optical properties, as discussed in the following section.

4.4
Optical Properties

In ZnSe$_{1-x}$Te$_x$ alloys, although Te is isovalent with Se, it does have different electronegativity and thus forms so-called isoelectronic centers, which can localize excitons and emit photoluminescence (PL). The microscopic configuration of these isoelectronic centers has been studied fairly extensively but still remains open to debate. Lately, single impurity states in bulk semiconductors, such as small isoelectronic centers, have been argued to be good quantum emitters [7, 8], similarly to quantum dots, while avoiding the material issues such as the interface between the quantum dots and their environment. We now note that, in contrast to the Te-related centers in the ZnSeTe alloys, there has been a prediction that, in the $In_xGa_{1-x}N$ system, indium atom clusters (isoelectronic centers) would localize excitons and behave essentially as quantum dots (QDs). It is obvious that it would be of great interest to compare the properties of small and large (QD-like) isoelectronic centers in the same material system.

Here we first briefly review the optical properties, specifically the PL, of Te isoelectronic centers in dilute bulk $ZnSe_{1-x}Te_x$ alloys. We then next discuss the δ-ZnSe:Te system, where the PL is still dominated by Te isoelectronic centers (see Ref. [9]) but the presence of QDs is indicated. Last, we focus on type-II QD-related PL in the δ^3-ZnSe:Te system and we show that isoelectronic centers and QDs coexist (see Ref. [12]), which makes this system desirable for studying and directly comparing the properties of small and large Te isoelectronic centers.

4.4.1
Dilute Bulk $ZnSe_{1-x}Te_x$ Alloys

The optical properties of dilute $ZnSe_{1-x}Te_x$ alloys have been studied fairly extensively, and it is generally agreed that the dominant low-temperature ($T \sim 10\,K$) PL is due to excitons bound to various Te clusters. More specifically, a PL band around 2.65 eV ("blue" band) usually dominates in $ZnSe_{1-x}Te_x$ with small Te concentrations (up to $x < 1.5\%$), while a band with a maximum at about 2.45–2.50 eV ("green" band) dominates in samples with larger Te concentrations. These bands are generally attributed to excitons bound to isoelectronic Te_{Se} atoms and/or clusters in these alloys. Some of the published identifications of these PL bands are summarized in Table 4.4 [9].

In Refs. [2, 32–34] the blue band was attributed to excitons localized at Te_2 complexes, and the green band to those localized at $Te_{n\geq2}$ complexes, or, more

Table 4.4 Assignments of Te-related PL in $ZnSe_{1-x}Te_x$ in the literature.

Ref.	PL energy (eV)	Assignment	PL energy (eV)	Assignment	Te (%)
2	2.61–2.63	Te_2 clusters	2.48–2.50	Te_3 clusters	<2
32	2.65	Te_2 clusters	2.50	Te_3 clusters	1–2
35	2.65	Te_1	2.50	Te_2 clusters	1
38	2.67	small Te clusters	2.50	small Te clusters	1
36	2.67	Te_1	2.48	$Te_{n\geq2}$ clusters	10–40
37	2.65	Te_1			1
33	2.65	$Te_{n\geq2}$ (distant)	2.45	Te_2 clusters	1–4

likely, $n = 3$ [2]. In Refs. [35–37] these two bands were attributed instead to self-trapped excitons at a single Te atom (Te$_1$) and at Te$_2$ complexes, respectively. In samples with very low (<1%) Te concentrations, lines at 2.75–2.784 eV were reported [33, 34, 37] and attributed either to free excitons [37], or to an exciton localized either at a Te atom [33] or in compositional fluctuations [34]. Lines with sharp longitudinal optical (LO) phonon replicas [2, 38] were attributed to resonant [34, 38] excitons or to an "impurity–Te" complex [2].

We note that, in most of the previous literature, there is little distinction between Te clusters (nearest neighbors) and other configurations. That there should be such a distinction, to some degree, was pointed out by Yang et al. [33]. These authors attributed the blue band with the maximum at 2.65 eV to excitons bound to two or more isolated Te atoms (i.e. pairs or triplets on non-nearest-neighbor sites). A further question is the origin of a peak at 2.75–2.78 eV. Yang et al. [33] attributed their 2.78 eV transition to an exciton localized at Te$_1$. A similar peak at 2.75 eV reported by Permogorov and Reznitsky [34] was attributed by them to free excitons localized by compositional disorder. However, it is unclear how such Te$_1$ bound excitons, which are expected to have a lower binding energy than those localized at Te$_2$ pairs, become dominant at higher temperatures as observed in Ref. [33]. As we will discuss further below, we feel that the assignment of this 2.78 eV peak to a free-to-bound transition or donor-bound excitons would be more appropriate, although further studies are needed to confirm this suggestion.

4.4.2
δ-ZnSe:Te

The PL spectra at $T = 10$ K from a typical δ-ZnSe:Te sample (A1) are shown in Fig. 4.4(a). There is a broad band with a maximum at around 2.64 eV, which is overlaid by sharp peaks on the high-energy wing. These sharp peaks can be grouped into three series [9], with the lines in each series separated by the ZnSe LO phonon (0.031 eV) energy: (a) 2.768, 2.736, 2.705 eV; (b) 2.758, 2.727, 2.696, 2.665, 2.634 eV; and (c) 2.751, 2.720, 2.689 eV. Figure 4.4(b) shows the PL spectra for this sample taken at different excitation intensities (for convenience, these curves and all subsequent ones are normalized to the corresponding maxima and arbitrarily shifted in the intensity scale). There is no substantial peak shift with excitation intensity, either for the band as a whole or for the sharp peaks at $T = 10$ K. At high excitation intensities one can also observe a small peak at about 2.798 eV with a shoulder on the high-energy side.

In Fig. 4.5(a) we plot the PL spectra from the same sample taken at the maximum excitation density, with Fig. 4.5(b) showing the PL obtained at the excitation intensity about two orders of magnitude lower. A further noteworthy feature of the spectra in Fig. 4.5 is that they indicate the presence of a peak at 2.61 eV, which can be best seen in the $T = 50$ K curves.

To understand the PL characteristics, we first note a strong similarity in many respects to the PL of bulk dilute ZnSe$_{1-x}$Te$_x$ alloys with relatively low Te content

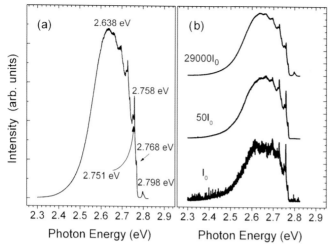

Fig. 4.4 (a) The photoluminescence spectrum ($T = 10\,K$) of sample A1. (b) Photoluminescence recorded at various laser excitation intensities (spectra arbitrarily shifted in the vertical direction for clarity).

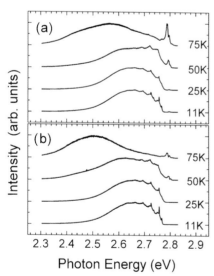

Fig. 4.5 Temperature evolution of photoluminescence of sample A1 (a) at the highest laser excitation intensity and (b) at an excitation intensity about two orders of magnitude lower.

mentioned above. Presumably, this blue band (2.64–2.65 eV) is due to the recombination of excitons localized at clusters [two or more nearest-neighbor (NN) Te atoms] or Te pairs, triplets, etc. (configurations other than NN), and its origin will become clear in the discussion that follows.

In examining the lowest-intensity curve in Fig. 4.4(b) and the data at 50 K (Fig. 4.5), we estimate that the blue band consists of a band with maximum at about 2.61 eV and contributions from the higher phonon replicas of the 2.758–2.768 eV transitions (see further discussion below). Note that some contribution from the green band (2.50 eV) is also possible, as seen at high temperatures (Fig. 4.5). It would be logical to assume that larger Te clusters would be responsible for deeper PL [39]; we thus ascribe the 2.60–2.61 eV band to excitons localized at Te_2 clusters. With this it follows that the green band is due to excitons localized at $Te_{n \geq 3}$ clusters, including Te-rich quantum dots, which will be discussed in more detail in conjunction with δ^3-ZnSe:Te. Interestingly, the observation of the green band shows that even in single δ layers there are enough large clusters ($Te_{n \geq 3}$) to give substantial PL, and, indeed, to form QDs (discussed below).

In view of the presence of large clusters, as shown above, it can be emphasized that it seems extremely likely that there is a distribution of cluster sizes.

As mentioned, these samples also show several series of lines fairly similar to each other, each with LO phonon replicas at 2.751, 2.758, and 2.768 eV. It seems plausible to assume that these lines are also due to excitonic-type centers. Moreover, further insight into these lines is provided by magneto-optical studies [Figs. 4.6(a) and (b)], which show that there is very little shift (<1 meV) of these lines even under a strong magnetic field (30 T). This indicates that these lines are due to very strongly bound excitons. We suggest that these sharp lines are due to excitons bound to non-nearest-neighbor pairs; it seems reasonable that there would be a number of such pairs at different separations. It is important to note, however, that recent calculations on GaP:N [40] showed that there is no unique relationship between neighbor separation and the energies of associated transitions; moreover, it has been suggested that at least one PL line in the GaP:N system belongs to triplets. Therefore, we believe that it is premature, at this time, to ascribe a given peak to a specific pair.

The assignment of these sharp lines to excitons bound to non-nearest-neighbor pairs is consistent with ascribing the 2.60–2.61 eV peak to Te_2 clusters, since one expects deeper levels for binding to nearest neighbors; our assignments are also consistent with observations and interpretations of the extensively studied case of N in GaP [41, 42].

A recent study on a single-monolayer Te-doped ZnSe embedded in the ZnSe matrix [8] argued that the sharp PL peaks observed between 2.6 and 2.8 eV are due to excitons bound to isolated Te_2 clusters, and that the inhomogeneity in the emission energy is due to the random strain field created by the Te atoms in the vicinity of isolated Te_2 clusters. We note, however, that the sharp peaks we have observed here have energy splittings larger than 0.6 meV as reported in Ref. [8]. This might indicate different origins from those observed in Ref. [8], and thus might be related

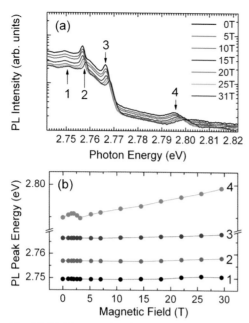

Fig. 4.6 (a) Magneto-photoluminescence of sample A1 at $T = 10\,K$. (b) The energy shifts of peaks 1, 2, 3 and 4 (indicated by arrows) as functions of the magnetic field.

to other Te configurations. On the other hand, if the sharp peaks we have observed were due to excitons bound to isolated Te_2 clusters, it could not account for the observation of the 2.60–2.61 eV peak in our case. It seems, therefore, that more studies, such as micro-magneto-PL, are needed for a conclusion to be reached.

Reverting to the PL shown in Fig. 4.4, we next note that, independently of the microscopic origin of the PL in this sample, the sharp lines do have quite strong LO phonon replicas and that, moreover, the PL also shows a peak at about 2.60–2.61 eV. This type of structure is typical of a relatively large Huang–Rhys factor (S) [43, 44]. Specifically, the situation that the no-phonon line (NPL) is weaker than phonon replicas is typical for centers with a strong Huang–Rhys coupling factor ($S > 1$) [44]. In the present case of Te in ZnSe, we have a hole–phonon interaction, which is usually very strong due to a large hole mass. We further note that for this type of PL from a single center, S can be obtained approximately from the position of the PL peak in relation to the NPL. In the present case, however, the observed broad peak with a maximum at 2.64 eV is a combination of different transitions, so several centers are involved. Nevertheless, using the fact that the approximate maximum of the series is close to the fourth phonon replica of the 2.758 eV transition [Fig. 4.4(a)], we can estimate $S \sim 4$.

As discussed, the sharp PL structure is due to non-nearest-neighbor pairs. Thus, it seems unlikely that isolated Te atoms are involved. This is consistent with the fact that no higher-energy lines are observed, since smaller entities are expected to give higher-energy PL. We note that, there cannot be single Te isoelectronic centers if a ZnSe:Te system forms alloys. Although there is no clear consensus on this, we suggest that our results favor the alloy view. We also note that formation of isoelectronic centers requires a large difference in either electronegativity [41, 42] or size [45] of the constituents [42, 44]; both differences appear nontrivial in ZnSe/ZnTe systems (difference in lattice constant is 7%), but there nevertheless seems to be no clear evidence for excitons associated with a single Te atom. We note that it has been established that the ZnS:Te system does have excitons localized at single Te atoms. However, one has to remember that (i) the electronegativity difference between Te and S is substantially larger than that between Te and Se and (ii) the ZnS–ZnTe system is not completely miscible (see e.g. Ref. [46]). Also, some theoretical first-principles calculations suggest that a single Te should not bind an exciton [47].

Finally, we would like to comment on the peak at 2.798 eV [shown for instance in Fig. 4.4(a)]. Magneto-optical studies [Fig. 4.6(b)] show that this peak exhibits a relatively large diamagnetic shift (~4 meV), indicating that this is due to either donor-bound excitons or a free-to-donor (FD) transition. As this peak dominates at high temperatures and high excitation intensities (Fig. 4.5), it seems more likely that it is due to an FD transition, since it is well known that FD transitions dominates ZnSe PL at high temperatures [48, 49]. Moreover, we believe that a shoulder on the high-energy side of this 2.798 eV line is due to free excitons that originate from either the undoped ZnSe layers or the buffer layer (undoped ZnSe). Indeed, the peak position of this shoulder shows significant energy shift under magnetic field (large diamagnetic shift), consistent with its free excitonic origin.

From the discussion above, it is clear that small Te_n isoelectronic centers dominate the PL properties of δ-ZnSe:Te. Moreover, though playing a smaller role, large Te clusters ($Te_{n\geq3}$) also contribute to the PL [Fig. 4.5(b)]. These $Te_{n\geq3}$ clusters already exhibit QD-like behaviors, as shown by the fact that the PL (green band) at $T = 80$ K shifts in energy as a function of the excitation intensity (Fig. 4.7), indicating a type-II band alignment between these clusters and the ZnSe matrix (see below). If more Te atoms are introduced into the δ layers, as in the case of δ^3-ZnSe:Te, one would expect an even more significant formation of larger Te clusters. It therefore naturally follows that more Te clusters are large enough to become Te-rich quantum dots, and this is indeed the case in δ^3-ZnSe:Te.

4.4.3
δ^3-ZnSe:Te

As discussed above in Section 4.3, the HRXRD studies suggest that the Te does not distribute uniformly in the plane of the δ layers, which, combined with the fact that Te exists only in submonolayer quantities, indicates that there is formation of Te-rich quantum dots (QDs) in these samples, especially δ^3-ZnSe:Te, where the Te concentration is relatively high (see Table 4.3). These ZnTe QDs are embedded in the ZnSe matrix, and the interface is characterized by a type-II band alignment [13]. Furthermore, the coexistence of type-II QDs and Te isoelectronic centers

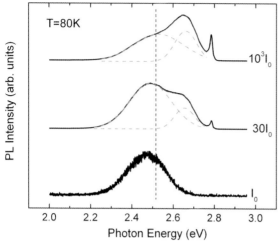

Fig. 4.7 Photoluminescence of a δ^3-ZnSe:Te sample (A2) at $T = 80$ K with various laser excitation intensities. The solid lines are fitting results with two Gaussian peaks (dashed lines). The vertical dashed line indicates the green band energy position under the maximum excitation intensity.

is also observed (see below), and this should provide an ideal platform to gain further insight into the evolution from few-atom systems (e.g. isoelectronic centers) to many-atom systems (e.g. QDs). Moreover, the size of these QDs can be controlled by changing the growth conditions.

4.4.3.1 Type-II Quantum Dots

The unique optical properties of type-II quantum structures arise from the spatial separation of electrons and holes. In the case of ZnTe/ZnSe quantum structures, photogenerated holes are localized in ZnTe while electrons are distributed in ZnSe (see Fig. 4.8). Due to the Coulomb force, electrons are attracted to the ZnTe/ZnSe interface, and, as a result, a triangular-shaped quantum well forms (Fig. 4.8). With increasing number of electrons and holes (e.g. by increasing the excitation intensity), the depth of this quantum well increases and the overlap between electron and hole wavefunctions becomes stronger. This leads to many interesting optical properties, one of which is the blueshift of the PL energy with increasing excitation intensity [50–52] as the level in the triangular-shaped quantum well is pushed higher. Indeed, as shown in Fig. 4.9, the green band PL from a typical δ^3-ZnSe:Te sample (B1) shifts toward higher energy by 42 meV when the excitation source is switched from a Xe lamp (weak excitation) to a He–Cd laser (strong excitation). More detailed studies of the excitation intensity dependence [12] show that, while the blue band (Fig. 4.10) exhibits very little shift, which is consistent with the isoelectronic bound exciton transition as shown above, the green band shifts more than 31 meV over four orders of magnitude of change of excitation intensity (Fig.

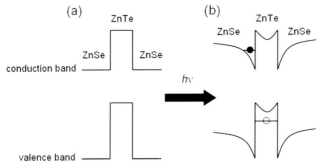

Fig. 4.8 Schematic band alignment between ZnSe and ZnTe (a) without laser excitation and (b) under laser excitation.

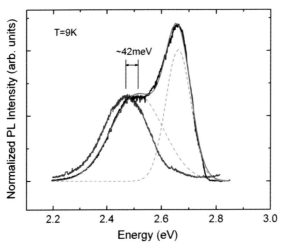

Fig. 4.9 Photoluminescence ($T = 9$ K) of a δ^3-ZnSe:Te sample (B1) under laser excitation (black line) and a Xe lamp excitation (blue line). The red line is the fitting result of two Gaussian peaks (green dashed lines) to the photoluminescence under laser excitation.

4.10). Moreover, the green PL peak position here follows the cube root of the excitation intensity (dashed line in Fig. 4.10), which has been predicted for type-II nanostructures [50].

The overlap between electron and hole wavefunctions as a function of the number of electrons and holes also leads to unique carrier dynamics, which are probed by time-resolved PL [for these measurements, a N_2 laser ($\lambda = 337$ nm) with pulsewidth of 4 ns was used] as shown below. The low-temperature ($T = 15$ K) PL decay, for the B1 sample, of the blue band (detected at 2.66 eV), and of the green band (detected at 2.46 eV), obtained under various excitation intensities, are

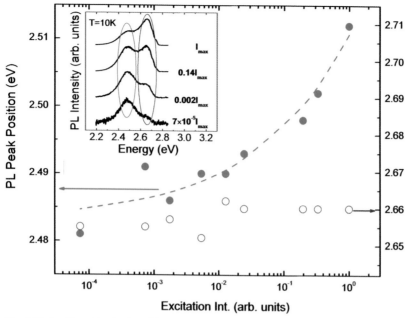

Fig. 4.10 Semilogarithmic plot of the green PL peak (the region marked in green in the inset) position (green solid circles) and the blue PL peak (the region marked in blue in the inset) position (blue open circles) at $T = 10$ K as a function of excitation intensity for a δ^3-ZnSe:Te sample (B1). The green dashed line is the result of fitting with $I^{1/3}$.

plotted in Figs. 4.11(b) and (c), respectively. The same data (detected at 2.66 eV) from a δ-ZnSe:Te sample (A1) is also shown, for comparison, in Fig. 4.11(a). It is clear that, while the blue PL decay remains a single exponential under various intensities, the green PL decay changes from a single exponential to nonexponential with increasing excitation intensity. Since the blue PL is due to isoelectronic bound excitons, single-exponential decay is expected. For the green PL, as the overlap of electron and hole wavefunctions is strongest right after the laser excitation pulse and becomes weaker over time [52], one would expect the PL to follow a nonexponential decay as observed here under strong excitation intensities. Under weak excitation intensities or a long time after the laser pulse, the overlap of wavefunctions approaches a constant as the numbers of electrons and holes decrease. This leads to the single-exponential decay observed here for the green PL. The characteristic PL decay time (τ_g for the green band and τ_b for the blue band) can be obtained under single-exponential decay conditions. We obtain $\tau_g \approx$ 86 ns; for the blue bands of δ-ZnSe:Te and δ^3-ZnSe:Te, we obtain $\tau_b \approx$ 30 ns and 38 ns, respectively. Note that $\tau_b \approx$ 16 ns at $T = 70$ K [solid circles in Fig. 4.12(d)] is close to the decay time of ~20 ns obtained at $T = 77$ K for the blue band in bulk ZnSeTe alloys [2]. As to the green band, there is a large discrepancy between

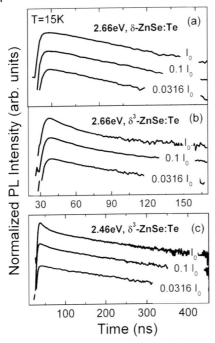

Fig. 4.11 Photoluminescence decay of (a) the blue band of a
δ-ZnSe:Te (A1), (b) the blue band of a δ³-ZnSe:Te (B1), and
(c) the green band of a δ³-ZnSe:Te (B1) under various laser
excitation intensities plotted in a semilogarithmic scale; the
detection energies are given in the graphs.

results for δ³-ZnSe:Te and bulk ZnSeTe alloys. Indeed, $\tau_g \approx 122$ ns obtained at $T = 70$ K [open circles in Fig. 4.12(d), see also below] for δ³-ZnSe:Te is much larger than that of alloys, which is ~35–40 ns at $T = 77$ K [2]. The similarity between the blue PL decay time here and in bulk alloys is expected as the PL originates from isoelectronic bound excitons. On the other hand, the spatial separation of electrons and holes in type-II QDs, hence the weaker overlap of electron and hole wavefunctions [52] compared with that in bulk alloys, leads to the long decay time of the green PL as observed.

The Coulomb energy between electrons and holes in type-II QDs can be extracted by considering the temperature dependence of τ of the green band in Fig. 4.12(d). Specifically, while the blue band PL decay time decreases monotonically with increasing temperature, the green PL decay time even increases with increasing temperature up to $T = 115$ K. An increase in τ can be readily understood for type-II QDs: as the temperature rises, the weakly bound electrons are ionized, and as a result are no longer bound to the strongly localized (confined) holes for an increasing fraction of their lifetimes, which in turn will lengthen the PL decay time. We note that such an explanation is valid provided that the nonradiative processes are

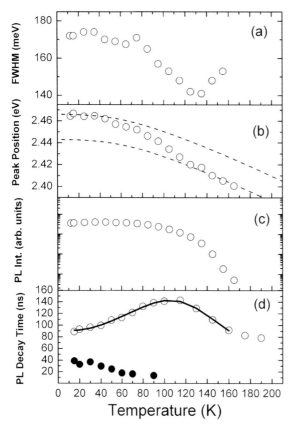

Fig. 4.12 (a) The FWHM, (b) the PL peak position (open circles), (c) the integrated PL intensity, and (d) the PL decay time (open circles) of the green band of a δ^3-ZnSe:Te (B1) as functions of temperature. Also, for comparison, the ZnSe bandgap temperature dependence (dashed lines), and the blue band PL decay time (solid circles) of a δ^3-ZnSe:Te (B1) are plotted in (b) and (d), respectively. The solid line in (d) is the fitting result using Eq. (4.4).

negligible, which is confirmed by the almost constant PL intensity in the same temperature region [Fig. 4.12(c), see also below]. Interestingly, this type of behavior was also observed for certain systems with isoelectronic centers (ZnTe:O [53] and GaAs$_{1-x}$P$_x$:N for some x values [54]) due to a similar effect. In Ref. [53], the temperature dependence of τ was fitted with the formula:

$$\tau_r = \tau^*/[1 - C\exp(-\varepsilon_{e-h}/kT)] \tag{4.3}$$

where τ^* is the decay time at $T = 0$ K, C is a constant, ε_{e-h} is a characteristic energy that is of the order of the electron–hole (e–h) binding energy, and k is the Boltzmann constant. With increasing temperature, τ does decrease; this is presumably due to nonradiative processes, and we use the relation

$$\tau^{-1} = \tau_r^{-1} + \tau_{nr}^{-1} \exp(-\varepsilon_{nr}/kT) \qquad (4.4)$$

where τ_{nr} is the characteristic nonradiative decay time and ε_{nr} is the activation energy of the nonradiative process. In Fig. 4.12(d) the solid line is the fitting result, and it yields $\varepsilon_{nr} \approx 8\,meV$, which is significantly lower than the free exciton binding energy in either ZnSe (~20 meV [55]) or ZnTe (~13 meV [55]). Such a low e–h binding energy is indeed expected for type-II QDs due to the spatial separation of electrons and holes [56].

4.4.3.2 Coexistence of Type-II Quantum Dots and Isoelectronic Centers

It seems likely that the green band shown here mostly originates from type-II ZnTe/ZnSe QDs and given the width of this band, it is possible that other radiative recombination centers, including Te isoelectronic centers, might also contribute. Indeed, as shown in Fig. 4.13, the PL decay time is not constant across the green band; instead, it decreases toward the high-energy side of the peak. The multicenter nature of the green band is further confirmed by plotting the temperature dependence of the full width at half-maximum (FWHM), the peak position and the integrated intensity of the green band in Figs. 4.12(a)–(c) respectively. The FWHM stays constant up to $T = 80\,K$ and then decreases until $T = 120\,K$ by as much as 35 meV, where it starts increasing. The peak energy undergoes a redshift (with increasing T), by as much as 25 meV, relative to the ZnSe bandgap in the temperature region where the narrowing of the FWHM is observed. Such a narrowing of the linewidth is in contrast to an increase in the FWHM observed from a similar green band (whose origin is due to excitons

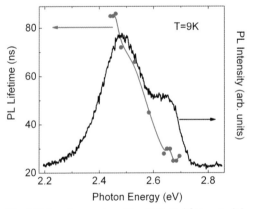

Fig. 4.13 Low-temperature ($T = 15\,K$) PL lifetime (solid circles) as a function of the detection energy across the PL spectrum of a δ^3-ZnSe:Te (B1). The solid line through the circles is a guide for the eye.

bound to Te$_3$ isoelectronic centers [2] in ZnSeTe alloys [35]). We note that the concurrence of the band narrowing and the redshift of the peak position is typical for an ensemble of QDs [57–62], which is explained by exciton transfer between QDs [57]. Therefore, we attribute this behavior to the existence of multiple centers associated with Te including QDs, and, quite possibly, Te isoelectronic centers (see also below). Also, within the same temperature region, the integrated PL intensity remains almost constant, as mentioned, which we have attributed to a negligible role of nonradiative processes; and this confirms that the narrowing of the FWHM and the redshift of the PL are due to carrier transfer within radiative recombination centers.

The multiple-center nature of the green PL is further elucidated by studying the PL excitation (PLE) and the absorption characteristics. The absorption and PLE spectra are shown in Figs. 4.14 and 4.15 (see also the inset). Both exhibit a similar, relatively broad, peak around 2.789 eV, with the absorption spectrum showing an additional, sharp, peak around 2.802 eV; we attribute the latter to the free exciton (FX) in the ZnSe buffer region, and the 2.789 eV peak to the free exciton from the spacer regions (ZnSe), whose energy is modified by the presence of Te due to some, minimal, Te diffusion and/or background Te presence. Using the formula given in Refs. [63–65], we estimated the Te concentration to be between 0.4% and 0.7%, consistent with the HRXRD results (Table 4.3). From the PLE detected at various energies across the PL spectrum (indicated by arrows 1–6 in Fig. 4.14), it is clear that the blue band energy region (arrow 1) as well as the high-energy part of the green band (arrow 2) are preferentially excited via the free exciton from the spacer regions, whereas the low-energy part (arrows 5 and 6) and the peak (arrow 4) of the green band are preferentially excited via band-to-band processes. Furthermore, for excitation with an energy in the tail of the absorption edge (Fig. 4.15),

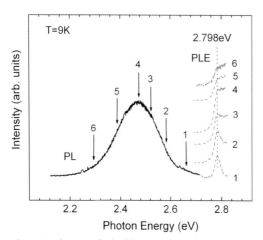

Fig. 4.14 The PLE (dashed lines) and PL spectrum of a δ^3-ZnSe:Te (B1) at $T = 10$ K. The PLE detection energies are indicated by the arrows (1–6) across the PL spectrum.

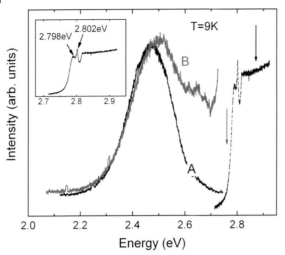

Fig. 4.15 PL spectra of a δ^3-ZnSe:Te (B1) obtained with the excitation energies indicated by the arrows A and B on the absorption spectrum.

the PL (curve B) shows an enhanced blue band and high-energy part of the green band, compared to excitation with above-bandgap energy (curve A).

To understand these results, we note that in ZnSeTe alloys, under the same excitation condition as that used to obtain curve B, the whole green band was enhanced [35]; therefore, it is obvious that the present green PL must have (at least) two different origins, one giving the low-energy part, and one giving the high-energy part. Furthermore, there are striking similarities in the PLE curves for our blue band as well as the high-energy side of our green band, and the PLE from alloys [35, 66]. Therefore, we attribute the high-energy part of our green band to excitons bound to Te isoelectronic centers, and the low-energy part of the green band to excitons associated with type-II QDs. This is consistent with the PL decay time profile across the green PL band (Fig. 4.13), as the isoelectronic center bound excitons have shorter lifetimes than type-II QD excitons [see e.g. Fig. 4.12(d)]. Moreover, the PL due to excitons bound to isoelectronic centers is enhanced under excitation by free excitons, due to direct capture by the isoelectronic centers; on the other hand, because of the spatially indirect nature of excitons associated with type-II QDs (spatially direct excitons cannot be directly captured by type-II QDs), the PL is strongly favored by the band-to-band excitation, where free carriers are generated.

Thus, PLE can be used to probe the formation of spatially indirect excitons; such excitons are directly linked to the formation of type-II QD band structures (energy barriers). As can be seen from Fig. 4.14, the free exciton peak at 2.789 eV (spatially direct exciton) on the PLE curve gradually disappears as the band-to-band excitation starts to dominate; this indicates that there is a smooth switch to PL from spatially indirect excitons with the change in detection energy (within the green band). This, in turn, suggests the smooth transition from isoelectronic centers to type-II QDs.

One example of a similar transition from isoelectronic centers to QDs is the possible formation of InN QDs in $In_xGa_{1-x}N$ alloys, for which it has been calculated [10] that as few as 13 In atoms are needed to localize the hole wavefunction, acting as an isoelectronic center, while a cluster of 201 In atoms can localize both the electron and the hole, and thus behave essentially as a QD. Moreover, it has been predicted [10] that, as the number of In atoms in a cluster increases, the properties (band structures) of such a cluster smoothly approach those of a QD. However, no experimental data are available for this system. Therefore, our Zn–Se–Te system, at present, is the only one on which experimental data for such transitions is available. We thus reiterate that the coexistence of small Te clusters acting as isoelectronic centers and large Te clusters behaving as QDs in this Zn–Se–Te multilayer system should provide an ideal platform to study this transition between few-atom systems and many-atom systems in more detail.

4.4.3.3 Controlling Quantum Dot Size by Varying Growth Conditions

The optical properties of these type-II ZnTe/ZnSe QDs can be further tuned by changing the growth conditions, which affect the size of the QDs. Specifically, larger QDs form with higher Te/Zn ratios (see Table 4.1) during multilayer deposition (see Ref. [67]). Here, we shall show that the optical properties change correspondingly as the QD size changes. For this purpose, we studied two samples, B2 and B3 (in addition to B1), which were grown with different Te/Zn ratios (see Table 4.1).

From Fig. 4.16, it is clear that the PL of B2, which was grown with a higher Te/Zn ratio, is redshifted compared to that of B3. In addition, the often observed blue band is absent in the PL of B2. These data suggest that (1) the size of the QDs in B2 is larger, resulting in the redshift of the PL, and (2) there are only very small quantities of small Te isoelectronic centers in B2.

The difference in the PL of B2 and B3 can be further seen in Figs. 4.17(a) and (b), where the spectrally resolved pulsed laser PL is shown for different time delays with respect to the laser pulse for samples B2 and B3, respectively. For sample B2, the PL spectra [Fig. 4.17(a)] at shorter delay times are dominated by a broad peak at 2.6 eV, which continuously shifts to the red with increasing delay time. The PL redshift with the delay time has been observed in type-II QDs and attributed to time-dependent band bending effect (see e.g. Ref. [68]). We note that the observed peak position at the shortest delay time is substantially higher than the peak energy usually observed for the green PL in δ-ZnSe:Te, but still below the blue PL associated with isoelectronic centers. We believe that this is due to the strong band bending under high excitation conditions of a pulsed laser; such an effect leads to the formation of higher energy levels for carriers (electrons in this case) and thus to the blueshift of the PL [50] compared to that observed from steady-state (lower-intensity) excitation sources. For sample B3, this type-II QD-related PL has a very weak intensity and appears only at delay times larger than 10 ns as a low-energy shoulder on the blue PL peak. Therefore, the shift of this type-II QD PL in sample

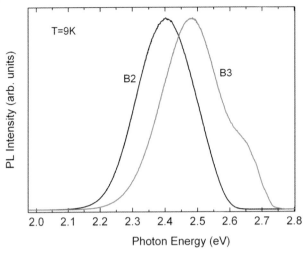

Fig. 4.16 PL spectra of samples B2 and B3, both of which are δ^3-ZnSe:Te but with different Te/Zn ratio during growth (see the text).

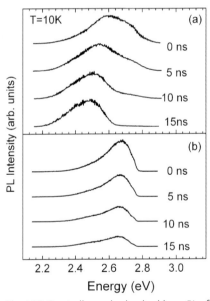

Fig. 4.17 Spectrally resolved pulsed laser PL of (a) sample B2 and (b) sample B3, with various delay times (indicated in the graphs) at $T = 10\,\text{K}$.

B3 is more difficult to observe. Nonetheless, considering that sample B2 has stronger type-II QD-related PL than sample B3 and that the shift of the PL energy also depends on the size of the type-II nanostructures [50], we suggest that there is formation of larger type-II QDs in sample B2.

To further confirm this, we studied the time decay characteristics of the PL. The results are shown in Figs. 4.18(a) and (b) for samples B2 and B3, respectively. Under high excitation intensities, the observed PL exhibits a nonexponential decay but approaches a single exponential as time elapses. As the excitation intensity decreases, the decay of the PL also approaches a single exponential, which is parallel to the one observed under the high excitation conditions at the longer times. These behaviors are consistent with what has been observed for the green PL in sample B1 (Fig. 4.11). The characteristic green PL decay times for both samples were estimated under single-exponential decay conditions, and this gives ~95 ns and ~86 ns at $T = 15$ K, for samples B2 and B3, respectively. The longer PL decay time for sample B2 suggests that the overlap between the electron and hole wavefunctions is weaker, consistent with the larger spatial separation of electrons and holes and thus the existence of larger type-II QDs in sample B2.

The existence of larger QDs in samples B2 and B3 is further substantiated by considering the Coulomb energy of spatially separated electrons and holes, which can be estimated by investigating the temperature dependence of the PL decay time, as shown above for the case of sample B1. Plotted in Figs. 4.19(a) and (b) are the decay times as functions of temperature for samples B2 and B3, respectively. It is clear that, for both samples, the PL decay time increases with increasing

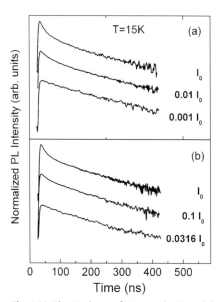

Fig. 4.18 The PL decay of (a) sample B2 and (b) sample B3 under various excitation intensities.

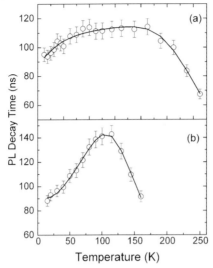

Fig. 4.19 PL decay time of (a) sample B2 and (b) sample B3 as a function of temperature. The circles represent the experimental data and the solid lines are the fitting results using Eq. (4.4).

temperature within certain temperature ranges, which has also been observed for sample B1. The solid lines in Figs. 4.19(a) and (b) are the fitting results obtained using Eq. (4.4), with $\varepsilon_{e-h} = 2\,\text{meV}$ for sample B2 and $\varepsilon_{e-h} = 8\,\text{meV}$ for sample B3. Both these values, as in the case of sample B1, are lower than the free exciton binding energy in either ZnSe or ZnTe. Furthermore, one would expect even lower exciton binding energies for larger type-II QDs due to the weaker overlap of electron and hole wavefunctions. Indeed, ε_{e-h} for sample B2 is less than that for sample B3, and this confirms that higher Te/Zn ratio leads to the formation of larger type-II QDs.

4.4.3.4 Magneto-PL of Type-II QDs: Aharonov–Bohm Effect and QD Size

Further evidence for the dot size dependence on the growth conditions is provided by magneto-PL studies [18], which were performed within the Faraday configuration in a magnetic field up to 31 T at $T = 4.2\,\text{K}$. The quantum dot size can be estimated based on the Aharonov–Bohm (AB) effect, which, in the case of type-II ZnTe/ZnSe quantum dots, arises from the interaction between the electron–hole dipole, with an electron circling around the quantum dots where the holes are located, and the external magnetic field. Theoretical studies on magneto-excitons (see e.g. Refs. [14–17]) predict that the emission energy of neutral excitons in type-II quantum dots will be affected by the Aharanov–Bohm [69] phase, which will result in changes in the exciton energy and the intensity of the excitonic PL. Specifically, the excitonic PL can be quenched in the magnetic field, which corresponds to the transition of the exciton angular momentum to a nonzero value

[14–17]. In an extreme simplification of a ring of zero width [14–17] the energy of an electron is

$$E_e = \frac{\hbar^2}{2m_e R^2}\left(l_e + \frac{\Phi}{\Phi_0}\right)^2 \tag{4.5}$$

where R is the radius of the ring, l_e is the electron angular momentum, Φ is the magnetic flux, and Φ_0 is the flux quantum ($= h/e$). Since, $\Phi = \pi R^2 B$, one expects the first oscillation at the magnetic field value $B = \Phi_0/2\pi R^2$.

A typical magneto-PL from sample B1, for several values of the magnetic field and two excitation intensities, is shown in Fig. 4.20. The PL intensity is a linear function of the excitation intensity for several orders of magnitude, suggesting that most of the QDs are not occupied. Therefore, it has been proposed [18] that an electron moves around a stack of QDs, one of which is occupied by a hole. The stacked cylindrical geometry nicely defines the ring-like trajectory for an electron, ensuring that the electron's wavefunction is "pushed" to the side of the dot, due to electron–electron interactions, independent of the stress in the system. It must be noted that for a single dot the electron will be located either above or below the dot, in the absence of strain, and, therefore, no AB signature is expected. Furthermore, the confinement potential for the hole and the barrier for the electron are relatively large, resulting in the electron wavefunction being pushed far from the

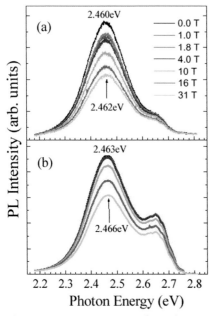

Fig. 4.20 Magneto-PL spectra of sample B1 under various magnetic fields: (a) with the laser excitation I_0, and (b) with the laser excitation $100I_0$.

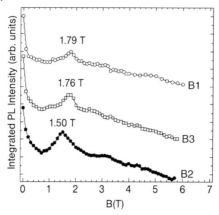

Fig. 4.21 Integrated magneto-PL intensity as a function of the magnetic field for samples B1, B2 and B3.

edge of the dots; this is expected to result in lower values of the magnetic field, where the AB effect is observed. In Fig. 4.21 the integrated PL intensity as a function of the applied magnetic field (sample B1) for the spectra obtained under the lowest excitation is plotted [18].

The overall integrated intensity decreases due to magnetic-field-induced carrier localization, as has been previously observed for type-II multiple-quantum-well structures (see e.g. Refs. [70, 71]). The major feature here is a strong peak at $B = 1.79 \pm 0.03$ T. The observed behavior is reversible. It is argued [18] that this peak is due to the AB phase, which is supported by calculations [see Fig. 4.22(a), first "arrow"], as described in Ref. [18]. The exciton binding energies in such stacks of ZnTeSe/ZnSe QDs computed within the single-band effective-mass model for both the conduction band and the valence band are in excellent agreement with experimental observations. Indeed, the computed binding energy of the exciton state at $B = 0$ T is 7.3 meV [Fig. 4.22(b)], and is in excellent agreement with the experimental value of 8 meV [12]. The calculations [18] have also shown that the lateral size of the QDs is 10–11 nm, whereas the maximum of the electron charge density is at 19–20 nm. The latter values can also be obtained using Eq. (4.5). Next, we show the effects of the Te fraction on the size of the dots. As clearly seen in Fig. 4.21, the magnetic field value required for the orbital momentum transition decreases with increasing Te/Zn ratio, indicating the presence of laterally larger quantum dots, which is in an agreement with the results of Ref. [67].

4.5
Summary

In summary, we have presented recent results on structural, optical, and magneto-photoluminescence investigations on Zn–Se–Te multilayer structures grown by the migration-enhanced epitaxy with submonolayer quantities of Te. Two types of

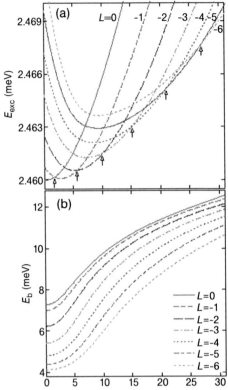

Fig. 4.22 (a) Dependence of the computed spin-down exciton energy levels on the magnetic field. (b) Binding energies of the lowest exciton states of different orbital momenta.

samples were considered (Fig. 4.1): δ^3-ZnSe:Te, where the Te/Zn cycle was repeated three times, and δ-ZnSe:Te, where only one Te/Zn cycle was used. Such growth procedures lead to the formation of Te-related centers with various sizes, probably due to the enhanced in-plane Te surface diffusion. This distribution in the size of Te-related centers gives rise to many interesting optical properties revealed by photoluminescence and magneto-photoluminescence. Specifically, it was shown the optical properties of δ^3-ZnSe:Te are dominated by type-II ZnTe/ZnSe QDs with Te isoelectronic centers also present, whereas the properties of δ-ZnSe:Te are similar to those observed from dilute bulk ZnSeTe alloys, with the QD characteristics appearing only at relatively high temperatures. Furthermore, it shown that there is a smooth transition between these two different centers, which is significant for understanding the scaling laws between many-atom systems (e.g. quantum dots) and few-atom systems (e.g. isoelectronic centers), and these multilayer structures should provide an ideal platform for further studies. The size of the quantum dots, which can be tuned by adjusting the Te/Zn flux ratio during growth, has been estimated using time-resolved PL and the Aharonov–Bohm oscillations in the PL intensity.

Acknowledgments

We would like to acknowledge support from DOE under grant number DE-FG02–05ER46219 as well as the contributions from Professor M. C. Tamargo (City College of CUNY) and her group for preparing the samples.

References

1 G. F. Neumark, R. M. Park, and J. M. DePuydt, *Physics Today* **1994**, 47, 26.

2 I. V. Akimova, A. M. Akhekyan, V. I. Kozlovski, Yu. V. Korostelin, and P. V. Shapkin, *Sov. Phys. Solid State* **1985**, 27, 1041.

3 I. L. Kuskovsky, Y. Gu, Y. Gong, H. F. Yan, J. Lau, I. C. Noyan, G. F. Neumark, O. Maksimov, X. Zhou, M. C. Tamargo, V. Volkov, Y. Zhu, and L. Wang, *Phys. Rev. B* **2006**, 73, 195306.

4 G. F. Neumark, I. L. Kuskovsky, and Y. Gong, "Doping aspects of Zn-based wide bandgap semiconductors", *Springer Handbook on Electronic and Photonic Materials*, Springer, Berlin, **2006**, p. 1845.

5 R. Brouri, A. Beveratos, J. P. Poizat, and P. Grangier, *Opt. Lett.* **2000**, 25, 1294.

6 C. Kurtsiefer, S. Mayer, P. Zarda, and H. Weinfurter, *Phys. Rev. Lett.* **2000**, 85, 290.

7 S. Strauf, P. Michler, M. Klude, D. Hommel, G. Bacher, and A. Forchel, *Phys. Rev. Lett.* **2002**, 89, 177403.

8 A. Muller, P. Bianucci, C. Piermarocchi, M. Fornari, I. C. Robin, R. Andre, and C. K. Shih, *Phys. Rev. B* **2006**, 73, 081306.

9 I. L. Kuskovsky, C. Tian, G. F. Neumark, J. E. Spanier, I. P. Herman, W. C. Lin, S. P. Guo, and M. C. Tamargo, *Phys. Rev. B* **2001**, 63, 155205.

10 L. W. Wang, *Phys. Rev. B* **2001**, 63, 245107.

11 K. Suzuki, U. Neukirch, J. Gutowski, N. Takojima, T. Sawada, and K. Imai, *J. Cryst. Growth* **1998**, 184/185, 882.

12 Y. Gu, I. L. Kuskovsky, M. van der Voort, G. F. Neumark, X. Zhou, and M. C. Tamargo, *Phys. Rev. B* **2005**, 71, 045340.

13 S. H. Wei and A. Zunger, *Phys. Rev. B* **1996**, 53, R10457.

14 A. B. Kalameitsev, V. M. Kovalev, A. O. Govorov, and *JETP Lett.* **1998**, 68, 669.

15 K. L. Janssens, B. Partoens, and F. M. Peeters, *Phys. Rev. B* **2001**, 64, 155324.

16 A. O. Govorov, S. E. Ulloa, K. Karrai, and R. J. Warburton, *Phys. Rev. B* **2002**, 66, 081309.

17 J. I. Climente, J. Planelles, and W. Jaskolski, *Phys. Rev. B* **2003**, 68, 075307.

18 I. L. Kuskovsky, W. MacDonald, A. O. Govorov, L. Muorokh, X. Wei, M. C. Tamargo, M. Tadic, and F. M. Peeters, cond-mat/0606752, *phys. Rev. B*, in press.

19 M. C. Kuo, C. S. Yang, P. Y. Tseng, J. Lee, J. L. Shen, W. C. Chou, Y. T. Shih, C. T. Ku, M. C. Lee, and W. K. J. Chen, *J. Cryst. Growth* **2002**, 242, 533.

20 D. Li and M. D. Pashley, *J. Vac. Sci. Technol. B* **1994**, 12, 2547.

21 Y. Gong, F. Y. Hanfei, I. L. Kuskovsky, Y. Gu, I. C. Noyan, G. F. Neumark, and M. C. Tamargo, *J. Appl. Phys.* **2006**, 99, 064913.

22 D. K. Bowen and B. K. Tanner, *High Resolution X-ray Diffractometry and Topography*, Taylor & Francis, London, **1998**.

23 N. N. Faleev, K. Pavlov, M. Tabuchi, and Y. Takeda, *Jpn. J. Appl. Phys.* **1999**, 38, 818.

24 M. Korn, M. Li, S. Tiong-Palisoc, M. Rauch, and W. Faschinger, *Phys. Rev. B* **1999**, 59, 10670.

25 P. F. Fewster, *X-ray Scattering from Semiconductors*, Imperial College Press, London, **2003**.

26 Y. Gong, H. F. Yan, I. L. Kuskovsky, I. C. Noyan, G. F. Neumark, and M. C. Tamargo, private communication.

27 S. Takagi, *Acta Cryst.* **1962**, 15, 1311.

28 D. Taupin, *Bull. Soc. Fr. Mineral. Crystallogr.* **1964**, 87, 469.

29 At a given depth z_j for an infinitely thin layer dz, X_j is the ratio of the diffracted wave amplitude $D_h(z_j)$ to the incident wave amplitude $D_0(z_j)$.

30 M. A. G. Halliwell, M. H. Lyons, and M. J. Hill, *J. Cryst. Growth* **1984**, 68, 523.

31 The imaginary component of χ_0 is negative due to the chosen waveform of $\exp(-i2\pi\mathbf{kr})$.

32 A. Yu. Naumov, S. A. Permogorov, A. N. Reznitski, V. Ya. Zhula, V. A. Novozhilov, and G. T. Petrovski, *Sov. Phys. Solid State* **1987**, 29, 215.

33 C. S. Yang, D. Y. Hong, C. Y. Lin, W. C. Chou, C. S. Ro, W. Y. Uen, W. H. Lan, and S. L. Tu, *J. Appl. Phys.* **1998**, 83, 2555.

34 S. Permogorov and A. Reznitsky, *J. Lumin.* **1992**, 52, 201.

35 D. Lee, A. Mysyrowicz, A. V. Nurmikko, and B. J. Fitzpatrick, *Phys. Rev. Lett.* **1987**, 58, 1475.

36 T. Yao, M. Kato, J. J. Davies, and H. Tanino, *J. Cryst. Growth* **1988**, 86, 552.

37 C. D. Lee, H. K. Kim, H. L. Park, H. Chung, and S. K. Chang, *J. Lumin.* **1991**, 48 & 49, 116.

38 S. Permogorov, A. Reznitsky, A. Naumov, H. Stolz, and W. von der Osten, *J. Lumin.* **1988**, 40 & 41, 483.

39 It has been shown, at least for GaP:N, that nitrogen clustering results in lower energy levels. These configurations, however, are highly strained and they are likely to occur only in high-impurity concentration samples (P. R. C. Kent, private communication). Note that δ-doped samples "mimic" such a situation.

40 P. R. C. Kent and A. Zunger, *Appl. Phys. Lett.* **2001**, 79, 2339.

41 D. G. Thomas, J. J. Hopfield, and C. J. Frosch, *Phys. Rev. Lett.* **1965**, 15, 857.

42 J. J. Hopfield, D. G. Thomas, and R. T. Lynch, *Phys. Rev. Lett.* **1966**, 17, 312.

43 W. Czaja and A. Baldareschi, *Helv. Phys. Acta* **1977**, 50, 606.

44 P. J. Dean and D. C. Herbert, "Bound excitons in semiconductors", in *Excitons*, ed. C. Cho, Springer, Berlin, **1979**.

45 J. W. Allen, *J. Phys. C* **1971**, 4, 1936.

46 H. Hartman, R. Mach, and B. Selle, *Wide Gap II–VI Compounds as Electronic Materials*, North-Holland, Amsterdam, **1982**.

47 S. H. Sohn and Y. Hamakawa, *Phys. Rev. B* **1992**, 46, 9452.

48 K. A. Bowers, Z. Yu, K. J. Gosset, J. W. Cook, Jr., and J. F. Schetzina, *J. Electron. Mater.* **1994**, 23, 251.

49 We also observed free-to-donor transition in lightly to intermediately doped ZnSe:N, and its intensity increased with temperature, as observed here, and in Ref. [48]. From temperature quenching, the activation energy of this transition is about 24 meV. This exactly corresponds to the difference between the peak energy (2.798 eV) and the bandgap of ZnSe (2.822 eV) at low temperatures.

50 N. N. Ledentsov, J. Bohrer, M. Beer, F. Heinrichsdorff, M. Grundmann, D. Bimberg, S. V. Ivanov, B. Ya Meltser, S. V. Shaposhnikov, I. N. Yassievich, N. N. Faleev, P. S. Kopev, and Zh I. Alferov, *Phys. Rev. B* **1995**, 52, 14058.

51 E. R. Glaser, B. R. Bennett, B. V. Shanabrook, and R. Magno, *Appl. Phys. Lett.* **1996**, 68, 3614.

52 F. Hatami, M. Grundmann, N. N. Ledentsov, F. Heinrichsdorff, R. Heitz, J. Bohrer, D. Bimberg, S. S. Ruvimov, P. Werner, V. M. Ustinov, P. S. Kopev, and Zh I. Alferov, *Phys. Rev. B* **1998**, 57, 4635.

53 J. D. Cuthbert and D. G. Thomas, *Phys. Rev.* **1967**, 154, 763.

54 J. A. Kash, J. H. Collet, D. J. Wolford, and J. Thompson, *Phys. Rev. B* **1983**, 27, 2294.

55 P. Y. Yu and M. Cardona, *Fundamentals of Semiconductors*, 2nd edn, Springer, Berlin, **1999**, p. 617.

56 U. E. H. Laheld, F. B. Pedersen, and P. C. Hemmer, *Phys. Rev. B* **1995**, 52, 2697.

57 Z. Y. Xu, Z. D. Lu, X. P. Yang, Z. L. Yuan, B. Z. Zheng, J. Z. Xu, W. K. Ge, Y. Wang, J. Wang, and L. L. Chang, *Phys. Rev. B* **1996**, 54, 11528.

58 G. Karczewski, S. Mackowski, M. Kutrowski, T. Wojtowicz, and J. Kossut, *Appl. Phys. Lett.* **1999**, 74, 3011.

59 S. Sanguinetti, M. Henini, M. Grassi Alessi, M. Capizzi, P. Frigeri, and S. Franchi, *Phys. Rev. B* **1999**, 60, 8276.

60 Motlan and E. M. Goldys, *Appl. Phys. Lett.* **2001**, 79, 2976.

61 B. Wang and S.-J. Chua, *Appl. Phys. Lett.* **2001**, 78, 628.

62 T. Passow, K. Leonardi, H. Heinke, D. Hommel, D. Litvinov, A. Rosenauer, D. Gerthsen, J. Seufert, G. Bacher, and A. Forchel, *J. Appl. Phys.* **2002**, 92, 6546.

63 A. Ebina, M. Yamamoto, and T. Takahashi, *Phys. Rev. B* **1972**, 6, 3786.

64 A. Yu. Naumov, S. A. Permogorov, T. B. Popova, A. N. Reznitsky, V. Ya. Zhula, V. A. Novozhilov, and N. N. Spendiarov, *Sov. Phys. Semicond.* **1987**, 21, 213.

65 M. J. S. P. Brasil, R. E. Nahory, F. S. Turco-Sandroff, H. L. Gilchrist, and R. J. Martin, *Appl. Phys. Lett.* **1991**, 58, 2509.

66 K. Dhese, J. Goodwin, W. E. Hagston, J. E. Nicholls, J. J. Davis, B. Cockayne, and P. J. Wright, *Semicond. Sci. Technol.* **1992**, 7, 1210.

67 Y. Gu, I. L. Kuskovsky, M. van der Voort, G. F. Neumark, X. Zhou, M. Munoz, and M. C. Tamargo, *Phys. Stat. Sol. (b)* **2004**, 241, 515.

68 A. A. Maksimov, S. V. Zaitsev, I. I. Tartakovskii, V. D. Kulakovskii, D. R. Yakovlev, W. Ossau, M. Keim, G. Reuscher, A. Waag, and G. Landwehr, *Appl. Phys. Lett.* **1999**, 75, 1231.

69 Y. Aharonov and D. Bohm, *Phys. Rev.* **1959**, 115, 485.

70 A. Truby, M. Potemski, and R. Planel, *Solid-State Electron.* **1996**, 40, 139.

71 M. Haetty, M. Salib, A. Petrou, T. Schmiedel, M. Dutta, J. Pamulapati, P. G. Newman, and K. K. Bajaj, *Phys. Rev. B* **1997**, 56, 12364.

5
Optical Properties of ZnO Alloys

John Muth and Andrei Osinsky

5.1
Introduction

The extremely efficient and bright luminescence from ZnO has been known for many years [1]. In the late 1950s and early 1960s, ZnO served as a model system for understanding the wurtzite semiconductor valence-band structure [2, 3] and also for observing the properties of excitons [4, 5]. It was also one of the first semiconductors investigated for optical [6] and electron-beam pumped lasers [7]. In the late 1990s, further studies have confirmed extremely bright photoluminescence and the ability of ZnO to lase as thin films [8], as random crystallites [9], and or in nanostructured forms [10]. As a light emitter in commercial applications, it has also been used successfully as a phosphor [11].

However, as an optoelectronic material, the lack of p-type doping has limited the development of ZnO p–n junctions [12–16]. This and the success of other semiconductor systems in the red and near-infrared portion of the spectrum throughout the 1970s and 1980s limited the investigation of ZnO for light-emitting diodes (LEDs). In the 1990s, the success of gallium nitride compounds for blue and green light emitters, as well as the many similarities between GaN and ZnO systems that are summarized in Table 5.1, led to a resurgence of interest in ZnO light emitters. It is hoped by many researchers that, like gallium nitride, a reasonably efficient method for p-type doping will be found [17, 18] such that ZnO homojunctions can be made [19] or that efficient hybrid devices such as those based on p-type GaN/n-type ZnO heterojunctions can be successfully grown and fabricated [15, 20–22]. If so, many anticipate that numerous applications for ZnO LEDs will be found.

The advantages of ZnO include the extremely high efficiency of exciton recombination based on the high exciton binding energy of ~60 meV. The high binding energy also provides ZnO devices with the ability to maintain efficient radiative recombination at elevated temperatures. The relatively lower index of refraction compared with III–V and III–nitride materials permits more efficient extraction of the light from ZnO since the critical angle for total internal reflection is larger.

Wide Bandgap Light Emitting Materials and Devices. Edited by G. F. Neumark, I. L. Kuskovsky, and H. Jiang
Copyright © 2007 WILEY-VCH Verlag GmbH & Co. KGaA, Weinheim
ISBN: 978-3-527-40331-8

Table 5.1 Comparison of gallium nitride and zinc oxide physical parameters.

	GaN	*ZnO*
Crystal symmetry	wurtzite	wurtzite
Lattice constants, a, c (Angstrom)	3.189, 5.185	3.252, 5.313
Bandgap energy at 300 K (eV)	3.4	3.34
Valence-band structure	A–Γ_9, B–Γ_7, and C–Γ_7	A–Γ_9, B–Γ_7, and C–Γ_7
Electron effective mass (me)	0.22	0.24
Hole effective mass (mn)	~0.8	~0.8
Exciton binding energy (meV)	26	60
Exciton radius (nm)	~3.2	~1.7
LO phonon energy (meV)	91	72
Electron mobility ($cm^2V^{-1}s^{-1}$)	1350	300 (predicted) 205 (observed) often around 100–150
Thermal conductivity ($Wcm^{-1}\,{}^0K^{-1}$)	2.3	1.1
Ionicity of bond	0.5	0.616
Hardness (GPa)	15.5	5.0
Melting point (°C)	2500	1975
Density (gcm^{-3})	6.15	5.61
Index of refraction	2.33	1.99

It is also believed that the availability of large native substrates, the environmentally benign nature of ZnO, and the ability to easily etch ZnO structures will potentially lead to economies of scale that will make ZnO-based devices potentially cheaper to produce. However, while ZnO LEDs have been demonstrated, numerous challenges remain. The principal challenge facing ZnO is reliable p-type doping with sufficient carrier concentration and mobility that is stable over time and environmental conditions.

The purpose of this chapter is to introduce the optical properties of the ZnO material system and its alloys. The fundamental optical properties of ZnO and its alloys are described from the optoelectronic device perspective, with emphasis on the room-temperature optical properties of interest for the design of light emitters. Specifically, the index of refraction, absorption, and luminescence are presented as functions of wavelength. These properties are connected by the physics arising from the wurtzite crystal symmetry and high exciton binding energy. The wurtzite symmetry leads to a splitting of the valence bands and to a uniaxial index of refraction as well as the A, B, and C excitons. The hexagonal crystal structure also leads to spontaneous polarization and piezoelectric effects when the material undergoes strain. The strain can then lead to high local fields and band bending within the device structure [22, 23]. In hybrid devices based on GaN/ZnO, and MgZnO/ZnO or CdZnO/ZnO heterostructures, it will be important to understand these effects to perform the necessary bandgap and strain engineering for proper carrier confinement and efficient recombination. While some data on band offsets, and MgZnO/ZnO and CdZnO/ZnO heterostructures, are present in the literature, this chapter avoids the subject of heterostructures and focuses more on bulk properties.

The effects of electric fields on excitons in bulk ZnO are discussed in detail. This discussion is also generally applicable to other excitons with large binding energy such as those in GaN. Understanding the properties of excitons as a function of temperature and electric field (locally generated by defects, generated by strain or applied externally) is important since many of the arguments to pursue ZnO as a light emitter rely on the implications that arise from the material having a large exciton binding energy. The main implication of the high exciton binding energy of ~60 meV is that it implies efficient recombination of the electron and hole, by keeping the electron and hole spatially correlated. The high binding energy also prevents thermal dissociation and allows the electron–hole pair to remain bound in relatively high local electric fields and under high carrier injection conditions.

There has been substantial interest in the growth of nanostructured ZnO, nanoparticles, platelets and needles. However, observation of the quantum confinement effect of the excitons is more difficult in the ZnO system as compared to other material systems since the smaller Bohr radius requires the confining structure to be smaller [24]. Thus, we leave the subject of ZnO nanostructures untouched except to note that photopumped ZnO nanostructures often exhibit very bright photoluminescence.

The remainder of this chapter discusses alloys of ZnO, principally MgZnO and CdZnO, as a means of controlling the bandgap and for the formation of superlattices. Moreover, BeZnO is mentioned since it has recently been demonstrated as a possible alternative to MgZnO.

Thus, this chapter is relatively narrow in focus and ignores many areas of study that are emerging, such as the use of ZnO as a spintronic material, the growth and properties of ZnO bulk substrates, or the potential uses of ZnO materials such as InGaZnO compounds for transparent electronics.

5.2
Index of Refraction of ZnO

ZnO has been grown by a wide variety of techniques. Thermodynamically, ZnO favors the wurtzite (hexagonal) phase over cubic structures, although some have reported deposition of zincblende (cubic) ZnO on substrates with cubic symmetry. Like III–V compound semiconductors, the zinc cation is tetrahedrally coordinated with the oxygen anion via sp^3 covalent bonding; however, the II–VI bond is substantially more ionic in nature.

The majority of ZnO thin-film studies have been performed on *c*-axis oriented sapphire substrates (α-Al$_2$O$_3$). There is also interest in growing *a*-plane ZnO on *r*-plane sapphire. In addition to differences in material quality that arise from differences in lattice matching and growth conditions, the orientation of the wurtzite crystal structure on the sapphire substrate also influences the optical properties of the material in fundamental ways by changing the orientation of the index of refraction ellipsoid with respect to the plane of the wafer. To understand the relative orientations of the crystals and the relation to the index of refraction ellipsoid, consider Fig. 5.1. Figure 5.1(a) shows the crystal planes and Miller indices, Fig. 5.1(b) the planes of the crystal cut to specific planes, and Fig. 5.1(c) the index of refraction ellipsoid for a positive uniaxial crystal with the extraordinary index aligned to the z-axis.

When growing c-axis oriented ZnO on c-axis oriented sapphire, the c-axis of the ZnO is normal to the surface of the wafer. When growing a-plane ZnO on r-plane sapphire, the c-axis of the ZnO is parallel to the wafer surface, and the a_3-axis is normal to the wafer surface. For the c-axis ZnO material grown on c-axis sapphire, the transverse electric (TE) and transverse magnetic (TM) polarization modes will have different effective indices of refraction, but the index of refraction will remain constant regardless of azimuthal angle. However, for a-plane ZnO grown on r-plane sapphire, the refractive index of the TM polarized light will remain constant, but the refractive index of TE polarized light will vary with the azimuthal angle according to

$$\frac{1}{n^2(\theta)} = \frac{\cos^2\theta}{n_o^2} + \frac{\sin^2\theta}{n_e^2} \tag{5.1}$$

where n_o and n_e are the ordinary and extraordinary refractive indices corresponding to the index of refraction ellipsoid of the ZnO film. The refractive index of a-plane ZnO for the TM polarization is n_o.

The index of refraction and thickness of thin semiconductor films can be measured by a variety of methods including ellipsometry. However, ellipsometry is very difficult to perform on birefringent transparent films on transparent

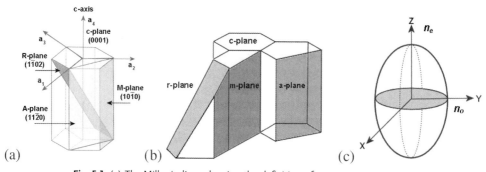

(a)　　　　　　　　　(b)　　　　　　　　　(c)

Fig. 5.1 (a) The Miller indices showing the definition of the commonly used planes of the wurtzite crystal structure for ZnO (and sapphire). (b) A solid figure showing the relationship between the different planes. (c) The index of refraction ellipsoid for a positive uniaxial crystal such as GaN or ZnO.

substrates. In the case of wide-bandgap semiconductor films on sapphire such as ZnO, a natural waveguide is formed since the refractive index of the ZnO is higher than that of the sapphire or air. Below the band edge where the absorption is low, prism coupling can be used to selectively excite the modes in the waveguide [25]. The prism coupling method requires a prism with a higher refractive index than the materials studied, as shown in Fig. 5.2(a). By bringing the prism in close proximity with the waveguide layer, the total internal reflection is frustrated. If the horizontal component of the propagation vector of the incident laser beam matches the propagation vector of the mode in the waveguide, very efficient coupling of energy occurs from the prism to the waveguide. This results in a sharp dip in the reflected light from the prism. Since the angle α at which the light is coupled into the waveguide can be measured precisely, and this is directly related to the wave number of the propagating mode the effective refractive index and thickness of the material can be measured very accurately especially if multiple modes can be excited as shown in Fig. 5.2(b).

Varying the incident wavelength of the laser in the prism coupling setup allows one to determine the refractive index as a function of wavelength. The refractive indices of TE_{min} and TE_{max} as a function of wavelength for a-plane sapphire are plotted in Fig. 5.3. The data were fitted by using the following Cauchy relation for refractive indices [26]:

$$n = A + \frac{B}{\lambda^2} + \frac{C}{\lambda^4} \tag{5.2}$$

where A, B, and C are constants. Table 5.2 gives the values for the constants obtained by performing the corresponding Cauchy fits. The fit between the curves and the data is in good agreement, indicating that Eq. (5.2) provides an accurate model of ordinary and extraordinary indices. Similarly, the refractive index data as a function of wavelength for pulsed laser deposition (PLD) grown c-axis oriented

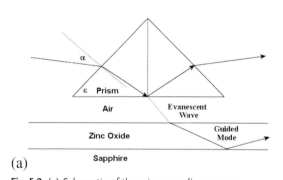

(a)

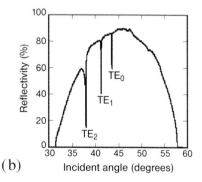

(b)

Fig. 5.2 (a) Schematic of the prism coupling process. (b) Reflection as a function of angle for an a-plane ZnO film showing three well-defined waveguide modes.

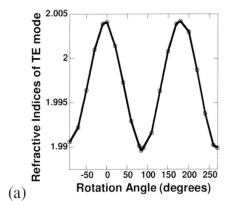

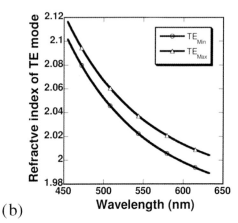

(a)   (b)

Fig. 5.3 (a) The angular dependence of *a*-plane sapphire as a function of the angular position of the wafer in the prism coupling apparatus at 632 nm. This is very different from the *c*-axis oriented ZnO where the TE mode remains constant as a function of the angular position of the wafer. In both cases the TM mode index is constant with the angular position of the wafer. (b) The refractive index as a function of wavelength for the TE_{max} and TE_{min} angular positions. It was also found that the TM mode corresponded to the TE_{min} values, as expected.

Table 5.2 Cauchy dispersion coefficients to describe the refractive index of *a*-plane ZnO grown on *r*-plane sapphire as a function of wavelength below the optical bandgap.

	A	*B* × 10³	*C* × 10⁹
TE_{max}	1.9723	−5.3564	7.2567
TE_{min} (also TM)	1.9555	−4.5323	7.1782

ZnO grown on *c*-axis sapphire substrates were measured and are shown in Table 5.3.

Above the bandgap where light is strongly absorbed, reflectivity or ellipsometry measurements can be used to obtain the refractive index. The complete dielectric function has also been computed theoretically, and has been experimentally determined using ellipsometry [27, 28] In the vicinity of the exciton resonance the index of refraction sharply increases due to the Kramers–Kronig relation. Using generalized ellipsometry, the dielectric function for ZnO platelets grown by vapor-phase transport has been measured, as shown in Fig. 5.4(a). One also finds that, with the addition of Mg, the refractive index of the MgZnO alloy is significantly reduced, since the refractive index (632 nm) of ZnO is about 2.0, while that of the wider-bandgap MgO is about 1.73. The ordinary and extraordinary refractive indices for

Table 5.3 Cauchy dispersion coefficients to describe the extraordinary and ordinary refractive indices of c-axis ZnO on c-plane sapphire below the optical bandgap.

	A	$B \times 10^3$	$C \times 10^9$
n_o	1.9484	−1.6148	6.8145
n_e	1.9574	3.9983	5.854

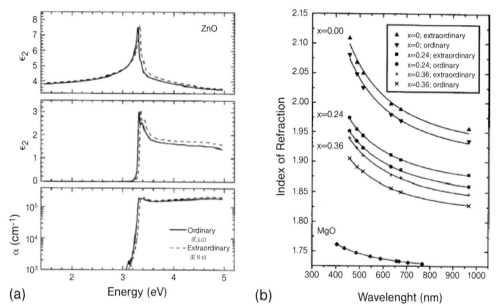

(a)

(b)

Fig. 5.4 (a) The real and imaginary parts of the complex dielectric function ε_1 and ε_2 and the absorption coefficient of ZnO grown by vapor transport and measured at room temperature. Reprinted from Jellison and Boatner [27]. (b) The extraordinary and ordinary refractive indices of MgZnO alloys [30].

ZnO, $Mg_{0.24}Zn_{0.76}O$, $Mg_{0.36}Zn_{0.64}O$, and MgO are shown in Fig. 5.4(b). Similarly, the refractive index for the cubic phase can be found in Ref. [29].

5.3
Excitonic Features of ZnO

The second major consequence of the crystal orientation is the interaction of polarized light absorption and subsequent emission. To understand this we need to consider the band structure of ZnO. The conduction band has an s-orbital like character with Γ_7 symmetry, while the valence bands are p-orbital in nature and are split into three bands by crystal-field and spin–orbit splitting. Unlike cubic III–V semiconductors, since the crystal has wurtzite symmetry, none of the valence bands are degenerate. While there has been substantial debate on the ordering of the ZnO valence bands, the present evidence points to a band ordering of Γ_9, Γ_7, Γ_7.

Transitions between the valence band and the conduction band are excitonic in nature, meaning that the Coulomb interaction between the electron and hole is strong, and that the electron–hole pair are spatially correlated and orbit each other like a hydrogen atom, as shown in Fig. 5.5(a). The convention is to label the three possible transitions A, B, and C, with B and C being the crystal-field and spin–orbit bands, respectively. In ZnO the effective-mass approximation and hydrogenic model provide a good description of the exciton. This allows one to find that the Bohr energy, or binding energy (Rydberg energy with $n = 1$) in electronvolts, is approximately

$$B_{\mathrm{ex}} = \frac{-13.6}{n^2} \frac{m_r^*}{m_0} \left(\frac{1}{\varepsilon}\right)^2 \tag{5.3}$$

with ε being the dielectric constant, m_r and m_0 the effective reduced mass of the exciton and the free mass of an electron, respectively, and n the quantum number for excited states of the exciton. Since the electron and hole are bound together by

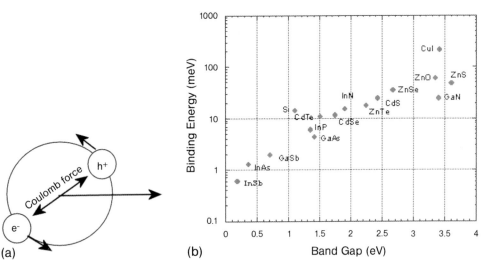

Fig. 5.5 (a) Hydrogenic exciton showing the electron and hole attracted by the Coulomb force with the center-of-mass motion of the exciton indicated by the arrow pointing to the right. (b) The exciton binding energy of a variety of materials plotted as a function of the bandgap energy of the materials. Note the log scale, and that the ZnO binding energy of 60 meV is significantly higher than kT at room temperature.

the Coulomb force, keeping them spatially correlated, the probability of recombination is very high, resulting in efficient emission.

Using group theory considerations and the assumption that the crystal is not strained, one can examine the probability of absorption and recombination under different polarization conditions, i.e. with the electric field E and wavenumber k being parallel or perpendicular to the crystal axis c. Thus one can consider three situations: σ-polarization with $E \perp c$ and $k \perp c$; π-polarization with $E \parallel c$ and $k \perp c$; and α-polarization with $E \perp c$ and $k \parallel c$. In the σ-polarization condition, all three transitions are allowed but the oscillator strength of the C exciton is much weaker. In the π-polarization condition, the C exciton is dominant, the B exciton is weak, and the A exciton transition is forbidden. Under the α-polarization condition all three transitions are visible.

The excitonic nature of the material also has a profound effect on the absorption and emission spectra [31]. In III–V semiconductors, the binding energy of excitons is substantially less than the thermal energy kT at room temperature of ~26 meV. Thus for these materials when considering the absorption mechanisms, one can use the parabolic band model and density of states very effectively, which for a direct bandgap semiconductor results in the absorption coefficient having a square-root dependence as a function of photon energy. However, as shown in Fig. 5.5(b), II–VI semiconductors can have substantially higher exciton binding energies well above kT at room temperature, with ZnO having a binding energy of ~60 meV. Thus when the Coulomb interaction is considered, one finds that the hydrogenic character of the exciton results in sharp absorption lines one Rydberg energy below the bandgap and also excited states of the exciton between the ground-state exciton resonance and the bandgap. These excited states are weaker in intensity, and merge into the continuum absorption after $n = 2$ or 3 and are not usually observable except at very low temperatures and in good unstrained material. Schematically, this is shown in Fig. 5.5(a). Note that each exciton A, B, or C can possibly have excited states as well, so the total absorption is the summation over all three valence bands. The excitonic nature of the absorption of ZnO is clearly seen in Fig. 5.6(b) [32]. Thus to find the bandgap of excitonic materials like ZnO one must consider the excitonic character, and to square the absorption spectra and linearly extrapolate to the energy axis as is commonly done with other materials is incorrect. Instead, the influence of the exciton on the band edge should be considered. In interpreting the excitonic structure of ZnO, in addition to considering the polarization dependence of the spectra, one also needs to consider the presence of electron–phonon replicas as additional peaks in the spectrum.

5.4
Electric Field Effects on Excitons

The Franz–Keldysh effect [33, 34] is used to describe the effects of an electric field on the absorption edge of an insulating semiconductor crystal. Using this theory the absorption edge takes the form $\exp(-E^{3/2}/f)$, where E is the energy measured

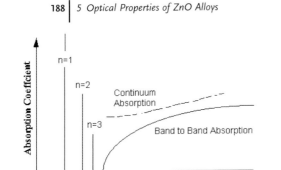

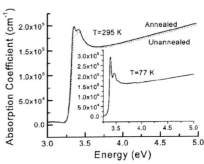

Fig. 5.6 (a) Schematic of the exciton absorption process with discrete lines and continuum absorption, as compared to the square-root-like dependence for direct bandgap semiconductors when a simple parabolic band model is used. (b) Absorption coefficient and exciton structure for an annealed (solid lines) and an unannealed (dotted line) sample at room temperature. The inset shows the absorption coefficient for the annealed samples at 77 K. The lower energy peak that is visible is the free exciton peak, while the second peak is the LO phonon replica of the free exciton. Modified and reprinted with permission from J. F. Muth, R. M. Kolbas, A. K. Sharma, S. Oktyabrsky, and J. Narayan, *J. Appl. Phys.* **1999**, 85, 7884. Copyright 1999, American Institute of Physics.

from the bandgap energy E_g and f is the applied electric field strength. However, the Franz–Keldysh effect was not intended to describe the effects of fields on the exciton since the absorption was assumed to arise from a single-electron band-to-band model and did not include the attractive force between electron and holes. Elliot [35] correctly described the exciton absorption spectrum including Coulomb effects, but it was not until Dow and Redfield [36, 37] that the effects of electric fields on the exciton absorption edge were computed. Figure 5.7(a) shows a calculation of the bound state of the exciton while under the influence of an electric field. For ease in calculation, the singularity of the Coulomb potential was removed. One finds that the majority of the wavefunction is still centered in the potential well, but the wavefunction is extended beyond the well following a form that asymptotically approaches that of an Airy function. Following the method of Dow and Redfield, the absorption due to excitons can be computed as shown in Fig. 5.7(b). Two effects on the absorption spectrum are readily apparent. The exciton resonance peak decreases strongly with field, and the absorption below the band edge increases.

The broadening of the absorption edge with electric field can be used to construct optical modulators [38, 39]. However, the broadening of excitons can also serve as a sensitive indicator of material quality, since local electric fields generated by atomic disorder in the crystal, or defects such as charged traps and dislocations or grain boundaries will also result in broadening of the exciton peak. It can be seen in Fig. 5.8(a) that annealing can be used to improve the crystal structure and that the absorption edge as well as the exciton resonance are sharper and more defined. Similarly in Fig. 5.8(b), one finds that the use of a buffer layer grown at low temperature and low pressure can be used to improve material quality.

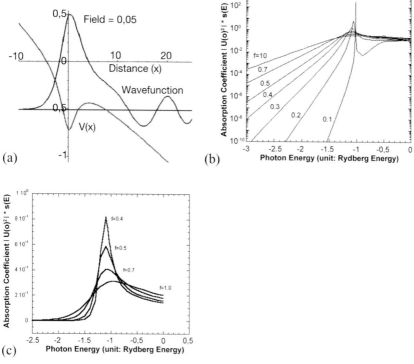

Fig. 5.7 (a) Effect of electric field on the Coulomb potential. For computational ease the Coulomb potential was modified to avoid the singularity at the origin. The wavefunction shows the bound state, and the extension of the wavefunction outside the well. (b) The Dow and Redfield model calculated with electron–phonon broadening neglected. Note the strong increase in absorption below the band edge. (c) The same Dow and Redfield model data plotted on a linear vertical scale. Note that only the energies below the band edge are plotted, and the continuum absorption that would cause increasing absorption above the band edge is not plotted.

5.5
Photoluminescence

There have been extensive photoluminescence studies on ZnO, especially at low temperatures, where donor and acceptor bound excitons can be used to identify the energy levels of impurities in the material. In this chapter, we will primarily focus on room-temperature photoluminescence (PL) and cathodoluminescence (CL), describe the basic features that can be expected of a good material, describe the green band emission associated with defects, and show examples of how annealing conditions can influence the emission from the material. A brief example of PL under high injection conditions will also be considered.

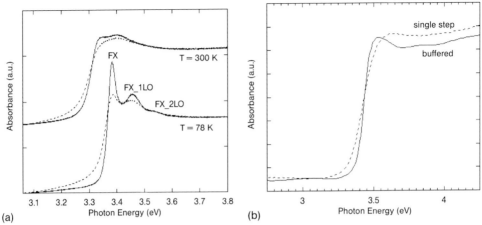

(a)

(b)

Fig. 5.8 (a) Room-temperature and 78 K absorption spectrum before and after annealing at 800 °C. (b) Significant improvements in the absorption edge can be obtained by using even a thin low-temperature buffer layer, before growing a thicker device layer. The improvement of the band edge is attributed to smaller magnitude microfields due to improved crystallinity.

The interest in ZnO for light emitters arises from a very strong band-edge PL. However, for light emitters it is important for the wavelength to be pure spectrally. In nonoptimized ZnO thin films one often finds, in addition to the near-ultraviolet emission, a broad green defect band that arises from defects. As an example of a sample with sharp band-edge emission, but also with green band emission, consider Fig. 5.9(a), showing the nonoptimized ZnO film grown by pulsed electron beam deposition. In Fig. 5.9(b), ZnO grown by PLD under optimized conditions shows that the green band emission is greatly suppressed compared to the band edge. One can also investigate the effects of stoichiometry on the green band luminescence by performing growth under different oxygen partial pressures, or as shown in Fig. 5.9(c), by post-annealing the films. We have found in general that the improvements in luminescence by post-growth annealing the films depend on the buffer layer and the growth conditions of the original film. However, in general by post-annealing in oxygen one finds that the green band PL or CL increases.

Under high excitation conditions, one finds that ZnO makes an efficient gain medium, and that stimulated emission and lasing action can be achieved. However, the exciton recombination dynamics can be complicated. In very pure material, exciton–exciton scattering can be observed, and at low temperatures bi-exciton recombination can be observed. However, at room temperature in typical materials, the PL as a function of pump power intensity often behaves as shown in Fig. 5.10.

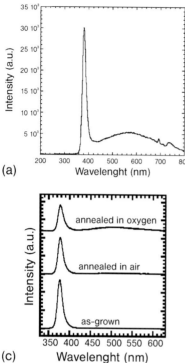

Fig. 5.9 (a) Cathodoluminescence of ZnO grown by pulsed electron-beam deposition, with an excess of oxygen. Note the sharp, strong band-edge luminescence, and the broad green band defect luminescence centered at 550 nm. Reprinted from H. L. Porter, C. Mion, A. L. Cai, X. Zhang, and J. F. Muth, *Mater. Sci. Eng. B – Solid State Mater.* *Adv. Technol.* **2005**, 119, 210.
(b) Photoluminescence of ZnO grown under optimized conditions by pulsed laser deposition. Note that the ratio of the band edge to defect luminescence is very high.
(c) One effect of annealing in oxygen can be an increase in green band photoluminescence.

5.6
ZnO Alloys

Bandgap engineering and the ability to form heterostructures is crucial for carrier and optical confinement in optoelectronic devices such as LEDs and lasers. Mg [40, 41] and Be [42] have been shown to increase the bandgap, while Cd [43] has been shown to decrease the bandgap.

The solid solubility limit of Mg in ZnO is about 2%. However, using nonequilibrium deposition techniques like molecular-beam epitaxy (MBE) or PLD, $Mg_xZn_{1-x}O$ with $x \sim 0.36$ can be grown while maintaining the wurtzite crystal structure of ZnO. Similarly while the solubility of Zn in MgO is low, using nonequilibrium growth techniques the alloys retain the cubic structure of MgO for $x > \sim 80\%$. These cubic MgZnO alloys can be grown on cubic substrates such as $\langle 100 \rangle$ silicon. In

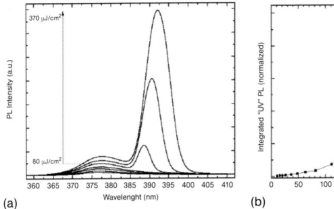

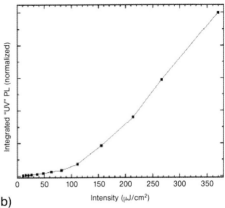

(a)   (b)

Fig. 5.10 (a) Photoluminescence spectrum for ZnO thin film excited by a femtosecond ultraviolet laser. Initially at lower pump powers the emission is dominated by the free exciton at 378 nm. However, as the excitation intensity is increased, an electron–hole gas is formed and Coulomb interactions result in a redshift of the photoluminescence. Under these conditions, the origin of the photoluminescence is largely by stimulated emission. (b) As indicated, at approximately $80 \, \mu J \, cm^{-2}$ a superlinear increase in the emission indicates the threshold of the process.

general, phase separation and the formation of MgO crystallites occurs for Mg contents between 36% and 80%. One interesting aspect of the phase separation is that the bandgap of MgO is higher than that of the surrounding materials. Thus the optical properties of the surrounding material of a lower bandgap are perturbed, but, assuming the amount of phase segregation is small, the optical properties are not strongly affected. This is in contrast to materials where the lower bandgap phase separates out, and results in the formation of a quantum dot, or other strong spatial perturbations of the bandgap potential.

Examining the bandgap versus lattice constant diagram, one finds that this allows one to create alloys with bandgaps that range from 3.4 to 4.3 eV and from ~7 to 8 eV, as shown in Fig. 5.11. The interest in light-emitting diode applications is in the near-ultraviolet portion using MgZnO and the visible portion of the spectrum using CdZnO compounds. The high quality of MgZnO alloys is shown in the absorption and PL spectra shown in Fig. 5.12.

The use of MgZnO as an alloy also permits the formation of MgZnO/ZnO heterostructures and quantum wells [44]. As shown in Figs. 5.13(a) and (b), this can be used to control the wavelength of the emission, although in this specific set of samples there were a substantial number of stacking faults observed in the TEM of the quantum wells.

As an alternative to using MgZnO to increase the bandgap, recently Ryu et al. have synthesized BeZnO compounds [42, 45]. BeO has a hexagonal phase, so may not suffer the same phase segregation issue as MgZnO compounds. While beryllium and beryllium oxide naturally occur in nature, there is a well-documented

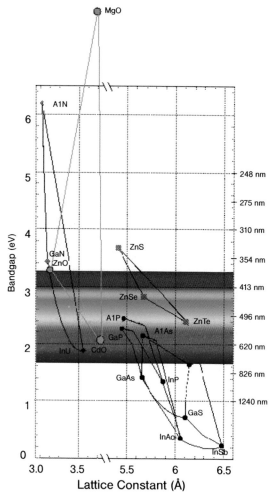

Fig. 5.11 Bandgap versus lattice constant diagram showing the relationship between the ZnO alloy system and other semiconductor compounds. Note that recent measurements indicate that the bandgap of InN is 0.7 eV.

health concern that people can become sensitized to beryllium and develop chronic beryllium disease [46]. Thus there are exposure regulations for processes where beryllium or beryllium oxide dust is formed and inhaled, but in solid form beryllium is widely used in many applications. As shown in Fig. 5.14, the band edge of higher-percentage BeZnO is not very sharp and is probably indicative of poorer crystal quality for higher-percentage compounds. As work with BeZnO progresses, one can expect the material quality to improve significantly.

Alloying cadmium with ZnO results in alloys with bandgaps that decrease with increasing Cd composition. Addition of Cd extends the emission of ZnO alloys to

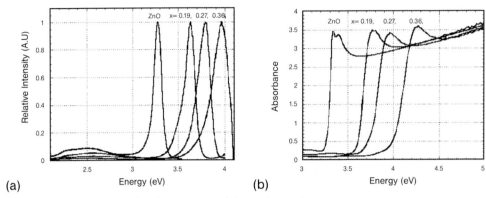

Fig. 5.12 (a) Photoluminescence of ZnO and MgZnO compounds. (b) Absorbance of spectra for MgZnO alloys. Note the exciton character of the band edge [41].

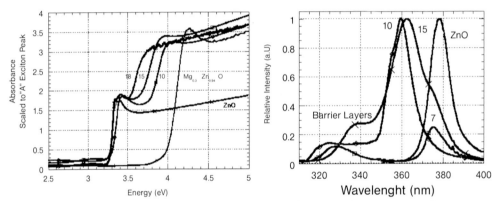

Fig. 5.13 (a) Absorption from MgZnO/ZnO quantum wells, where 18, 15, and 10 refer to the growth time (seconds) of ZnO. The $Mg_{0.34}Zn_{0.66}O$ and ZnO absorption curves are shown for reference. The shift of the MgZnO band edge with increasing well width indicates that diffusion is lowering the bandgap of the barrier layers. (b) The photoluminescence of MgZnO/ZnO quantum wells, for 7 s, 10 s, and 15 s growth times for the lower bandgap ZnO layer. The intensities of the 10 s and 15 s quantum wells are normalized to the ZnO peak intensity, but actually are substantially brighter than the ZnO photoluminescence, while the 7 s ZnO growth was lower in intensity, as shown. Reprinted from J. F. Muth, C. W. Teng, A. K. Sharma, A. Kvit, R. M. Kolbas, and J. Narayan, in *Laser–Solid Interactions for Materials Processing*, eds. D. Kumar, D. P. Norton, C. B. Lee, K. Ebihara, and X. Xi, *Mater. Res. Soc. Proc.* **2001**, 617.

the visible range. This is due to the lower bandgap of CdO, which is 2.3 eV at room temperature [12]. Moreover, Zn and Cd atoms have similar ionic radii [8], making the formation of alloys more straightforward. While MgO and ZnO are relatively nontoxic, Cd, a heavy metal, potentially poses some environmental and health risks.

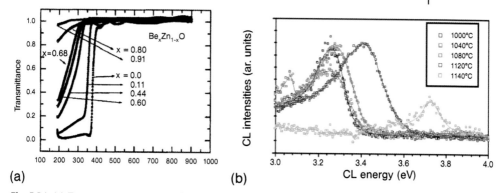

Fig. 5.14 (a) Transmission spectrum of BeZnO alloy measured at room temperature [42]. (b) Cathodoluminescence spectra for BeZnO thin films as a function of evaporation cell temperature [45].

Technologically, the difficulty with Cd is maintaining good stoichiometry and minimizing fluctuations in composition [47]. Fluctuations in Cd concentration can cause spatial fluctuations in the bandgap potential. Potentially these fluctuations can result in improved radiative recombination, as has been postulated to occur in the InGaN material system with composition inhomogeneities; although this effect should be less since the exciton binding energy is larger than the nitride material system. CdZnO alloys have been successfully grown by techniques on a-plane sapphire substrates [48, 49] and by radiofrequency (RF) plasma MBE on GaN/sapphire templates [50, 51]. RF plasma MBE methods have been especially successful in producing CdZnO films since in situ reflection high-energy electron diffraction (RHEED) monitoring allows monitoring of the crystal growth and balancing the stoichiometry.

Figure 5.15 shows $\theta - 2\theta$ X-ray diffraction (XRD) scans from MBE-grown $Cd_xZn_{1-x}O$ films with Cd mole fraction in the alloys ranging from 0.02 to 0.78 as measured by Rutherford Backscattering (RBS) and secondary-ion mass spectrometry (SIMS) techniques. The films were grown by RF-plasma MBE on GaN/sapphire templates using a thin ZnO buffer [50]. As can be seen in the figure, only the (0002) reflections from $Cd_xZn_{1-x}O$ films and the ZnO/GaN buffers appear. Taken in combination with the in situ RHEED data shown on the right-hand side of Fig. 5.15, this indicates that the $Cd_xZn_{1-x}O$ films grew with the epitaxial relationship $[11\bar{2}0](0001)_{CdZnO} \parallel [11\bar{2}0](0001)_{GaN}$ and have a wurtzite structure. As the Cd mole fraction in the $Cd_xZn_{1-x}O$ alloys increases, the (0002) peak position gradually shifts to smaller angles, indicating an increase in the $Cd_xZn_{1-x}O$ out-of-plane lattice constant, which is consistent with a study reported in [47]. It is important to note that no indication of CdO rocksalt phase was detected for $Cd_xZn_{1-x}O$ alloys with Cd mole fractions as high as 0.78. This suggests that phase segregation did not occur in the films, at least to the XRD measurement limit. As the Cd mole fraction increases in $Cd_xZn_{1-x}O$ alloys, the full width at half-maximum (FWHM) of the

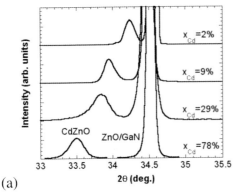

(a)

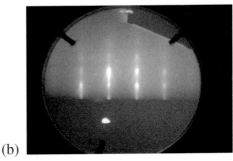

(b)

Fig. 5.15 (a) XRD $\theta - 2\theta$ scans of $Cd_xZn_{1-x}O$ films. No signature of rocksalt CdO is observed up to Cd mole fraction of 0.78. (b) In situ RHEED image from as-grown CdZnO surface.

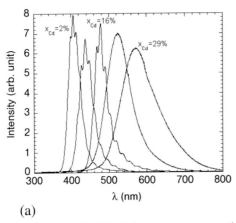

(a)

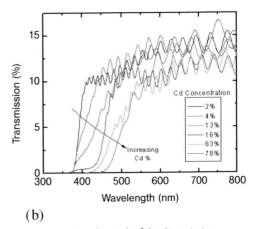

(b)

Fig. 5.16 (a) Room-temperature visible cathodoluminescence emission spectra from MBE-grown $Cd_xZn_{1-x}O$ ($0 < x < 0.29$) films. (b) Transmission spectrum of CdZnO thin films grown on ZnO/GaN templates. Comparison of bandgap obtained by measuring the peak of the CL emission intensity, and bandgap obtained by extrapolating the CdZnO absorption edge. Note the strong disagreement for higher Cd compositions, which is symptomatic of band-edge tailing and compositional fluctuations.

$Cd_xZn_{1-x}O$ (0002) peaks gradually broadens, which may be due to an increased defect density caused by a larger lattice mismatch.

Optical emission and bandgap properties of CdZnO alloys are of great interest for fabricating device heterostructures. Figure 5.16(a) shows five room-temperature (RT) CL spectra for $Cd_xZn_{1-x}O$ alloys with x mole fraction less than 0.3. Each spectrum has a unique single-peak emission band, and as the Cd mole fraction of

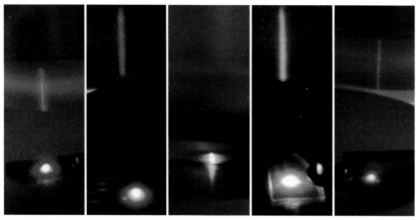

Fig. 5.17 Cathodoluminescence images for CdZnO films with emission ranging from 410 nm (right) to 580 nm (left). The streaks at the top of the photographs are an internal reflection within the CL system.

the CdZnO alloys increases from 0.02 to 0.29 the peak of the emission band redshifts from 410 nm to 580 nm. Very similar optical properties were measured for $Cd_xZn_{1-x}O$ alloys grown by the plasma enhanced (PE) metal–organic chemical vapor deposition (MOCVD) technique [48].

Figure 5.16(b) shows transmission spectra for CdZnO films with Cd mole fraction increasing from 0.02 to 0.78. The fundamental band edge shifts toward longer wavelength as the Cd mole fraction increases. Fringe patterns caused by optical interference in the films can be seen in the figure. Photographs of optical emission from 0.2 µm thick CdZnO films with various mole fractions of Cd are shown in Fig. 5.17 [51]. The images were obtained in a CL system with an electron-beam current set at 3 µA and accelerating voltage of 3 kV.

Photoluminescence excitation experiments were conducted with a 351 nm Ar laser operating at up to $1300 \, W \, cm^{-2}$ to investigate the nature of the visible light emission. The integrated light output intensity scales almost linearly with respect to the pumping power for CdZnO films with Cd mole fraction in the range of 0.02 to 0.29, which strongly suggests that the emission is associated with near-bandedge optical transitions rather than defect-induced transitions. For CdZnO film with 0.02 mole fraction of Cd, the FWHM of the peak is 35 nm, which is broader than for ZnO films (typically ZnO FWHM is ~14 nm at room temperature). As the Cd mole fraction increases to 0.29, the FWHM increases to 85 nm. This increase is consistent with the results from the XRD measurements and may again be due to a higher defect density caused by a larger lattice mismatch or composition fluctuations. The samples with Cd mole fraction above 0.3, i.e. $Cd_{0.63}Zn_{0.37}O$ and $Cd_{0.78}Zn_{0.22}O$, did not show cathodoluminescent emission. The optical bandgap for the film with high mole fraction of Cd was determined with transmission measurements.

Figure 5.18 summarizes the fundamental bandgap of CdZnO alloys plotted as a function of Cd mole fraction. Bandgap values were determined from the peak position of the cathodoluminescence emission bands for the films with Cd mole fraction up to 0.29 and from the optical transmission measurements for the films with Cd mole fraction up to 0.78. Note the presence of a strong bowing effect. Since ZnO and CdO have different crystal structures, the strong nonlinear relationship between CdZnO optical bandgap and Cd mole fraction is not unexpected. The transmission measurements for $Cd_{0.78}Zn_{0.22}O$ films indicate a bandgap of ~2.08 eV, which is significantly smaller than that of ZnO. The MBE-grown CdZnO films show slightly larger bandgaps for Cd mole fractions above 0.5 as compared to MOCVD-grown films [47]. This may be due to the different growth methods, substrate orientation, and measurement techniques used to determine the optical bandgaps and compositions of the films.

Uniform distribution of composition of ternary alloys can be explored by spatially resolved two-dimensional (2D) CL intensity mapping. CL mapping and spectral analysis were carried out for $Cd_{0.16}Zn_{0.84}O$ film at 77 K. The CL emission band for this film peaked at ~475 nm. The 2D CL map (imaging wavelength is 475 nm) shown in Fig. 5.19(a) features randomly distributed regions of ~1 μm in size of varying contrast without any clearly discernible boundaries separating them. The contrast between the darker and brighter regions is rather small. In Fig. 5.19(b), the CL spectra for a bright region and a dark region are superimposed for comparison. Both spectra have similar characteristics including the same peak wavelength, but the spectrum for the brighter spot has a slightly greater intensity. Significantly, this implies that large-scale phase segregation or Cd composition fluctuation does not take place in the $Cd_{0.16}Zn_{0.84}O$ film. The CL intensity fluctuations may arise as the result of defects.

Room- and low-temperature time-resolved photoluminescence (TRPL) measurements were performed on $Cd_{0.16}Zn_{0.84}O$ films grown on a ZnO substrate [50, 51].

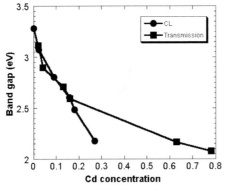

Fig. 5.18 Fundamental bandgap of CdZnO versus Cd mole fractions. Bandgap values extracted by CL are indicated by full circles, while those obtained from transmission measurements are denoted by full squares.

(a)

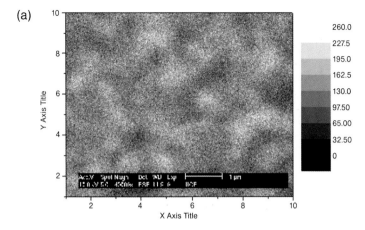

(b)

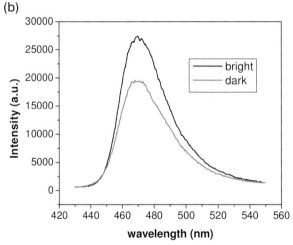

Fig. 5.19 (a) A spatially resolved 2D CL map and (b) CL spectra for bright and dark regions for $Cd_{0.16}Zn_{0.84}O$ film at 77 K.

The room-temperature spectra are shown in Fig. 5.20. The intensity decay (450 nm peak) consists of two exponential components. The first of these has a "short" lifetime of 21 ps while the other has a "long" lifetime of 49 ± 3 ps. The uncertainty in the second value is related to noise. The multiple exponents, which are directly related to lifetimes, tend to suggest that there are compositional microscale fluctuations present in the CdZnO film. The scale of the composition fluctuations is too small to be detected with the CL mapping shown in Fig. 5.19.

The 14 K TRPL results are shown in Fig. 5.21 for the same peak as in Fig. 5.20 (i.e., at 450 nm). The intensity decay is slower and nonexponential in this case, however. Alternative models for the intensity decay are the power-law decay given by

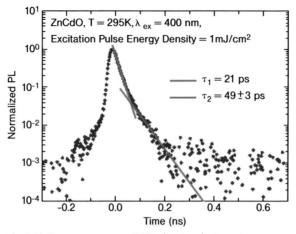

Fig. 5.20 Room-temperature TRPL showing the intensity decay of the 450 nm peak of a $Cd_{0.16}Zn_{0.84}O$ film.

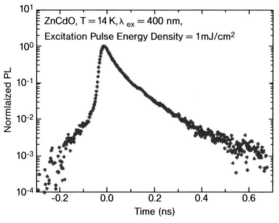

Fig. 5.21 TRPL peak intensity decay at 14 K. The sample is the same as for Fig. 5.20.

$$I \sim s^{t-b} \tag{5.4}$$

and the stretched-exponential decay given by

$$I \sim \exp[-(t/\tau_{eff})\beta] \tag{5.5}$$

where τ_{eff} is the effective lifetime and β is the stretched-exponential parameter.

In Figs. 5.22(a) and (b) the 14 K TRPL spectrum is fitted to the models given by Eqs. (5.4) and (5.5), respectively. The plot in Fig. 5.22(a) is not linear, indicating that the peak intensity decay does not follow a power law. On the other hand, the plot in Fig. 5.22(b) is linear, indicating that the intensity decay follows a stretched

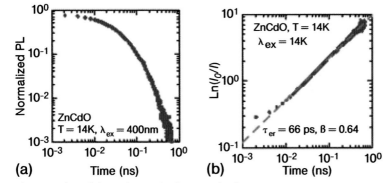

Fig. 5.22 Plots of the peak intensity decay results from Fig. 5.7 fitted to (a) the power law of Eq. (5.4) and (b) the stretched-exponential decay from Eq. (5.5).

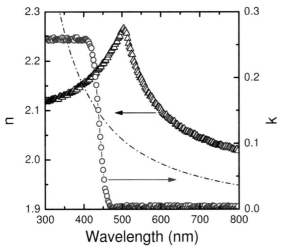

Fig. 5.23 Real (n) and imaginary (k) index of refraction values for a $Cd_{0.16}Zn_{0.84}O$ film as a function of wavelength. The values were determined through fitting of spectrophotometric data using a genetic algorithm. For comparison, the dashed line indicates the n value of ZnO as determined using a Sellmeier fit.

exponential, although the origin of this type of decay still needs to be investigated. This type of decay is similar to what has been seen in InGaN and amorphous silicon and is thought to be due to nonuniformity in the material. The same could be true for CdZnO.

Optical constants, such as n and k, were determined for a $Cd_{0.16}Zn_{0.84}O$ film from an experimental optical absorption curve. The values of n and k were extracted by a fitting algorithm that uses both Sellmeier and Forouhi–Bloomer relations. Figure 5.23 provides the calculated n and k values for a $Cd_{0.16}Zn_{0.84}O$ film in the 300–800 nm spectral range. For comparison, the n value for ZnO using a similar fit has been included in the graph, represented by the dashed line. The real index of refraction for $Cd_{0.16}Zn_{0.84}O$ is generally higher than that of ZnO, varying between 2.02 and 2.26 in the 300–800 nm spectral range. A correspondingly sharp drop in the k value for $Cd_{0.16}Zn_{0.84}O$ occurs near 440 nm. These results are very promising and show that it is possible to make MgZnO/CdZnO heterostructures with relatively large index contrast, since Mg and Cd can be alloyed with ZnO to decrease or increase the index of refraction, respectively.

Acknowledgments

John Muth would especially like to thank Drs. Xiao "Roy" Zhang, Ailing Cai, Hugh Porter, and Tim Teng, who as graduate students worked very hard taking and interpreting much of the data discussed in this chapter, Ajay Sharma and Jay Narayan for the ZnO and MgZnO crystal growth, and Henry Everitt's group with John Forman who took the photoluminescence data. John Muth would also like to thank the Office of Naval Research Young Investigator Program. Andrei Osinsky would like to thank Dr. Brian Hertog, for help with data analysis and preparation of the manuscript, Drs. Jianwei Dong and Junqing Xie for diligent efforts in crystal growth, Professors Winston Schoenfeld, Leonid Chernyak, and Alexander Cartwright for useful discussions and providing RBS, optical index, CL mapping and time-resolved PL results, and Professors Steve Pearton, David Look, and Alexander Mintairov for their insightful comments. Both authors would like to thank the Army Research Office and program manager Dr. Mike Gerhold, who has been instrumental in supporting this research.

References

1 U. Ozgur, Y. I. Alivov, C. Liu, A. Teke, M. A. Reshchikov, S. Dogan, V. Avrutin, S. J. Cho, and H. Morkoc, *J. Appl. Phys.* **2005**, 98, 041301.

2 J. J. Hopfield, *J. Phys. Chem. Solids* **1959**, 10, 110.

3 J. J. Hopfield, *J. Phys. Chem. Solids* **1960**, 15, 97.

4 R. E. Dietz, D. G. Thomas, and J. J. Hopfield, *J. Appl. Phys.* **1961**, 32, 2282.

5 D. G. Thomas and J. J. Hopfield, *Phys. Rev. Lett.* **1961**, 7, 316.

6 J. M. Hvam, *Solid State Commun.* **1973**, 12, 95.

7 F. H. Nicoll, *Appl. Phys. Lett.* **1966**, 9, 13.

8 D. M. Bagnall, Y. F. Chen, Z. Zhu, T. Yao, S. Koyama, M. Y. Shen, and T. Goto, *Appl. Phys. Lett.* **1997**, 70, 2230.

9 H. Cao, Y. G. Zhao, H. C. Ong, S. T. Ho, J. Y. Dai, J. Y. Wu, and R. P. H. Chang, *Appl. Phys. Lett.* **1998**, 73, 3656.

10 M. H. Huang, S. Mao, H. Feick, H. Q. Yan, Y. Y. Wu, H. Kind, E. Weber, R. Russo, and P. D. Yang, *Science* **2001**, 292, 1897.

11 K. Vanheusden, W. L. Warren, C. H. Seager, D. R. Tallant, J. A. Voigt, and B. E. Gnade, *J. Appl. Phys.* **1996**, 79, 7983.

12 S. J. Pearton, D. P. Norton, K. Ip, Y. W. Heo, and T. Steiner, *Prog. Mater. Sci.* **2005**, 50, 293.

13 D. C. Look, *Mater. Sci. Eng. B – Solid State Mater. Adv. Technol.* **2001**, 80, 383.

14 D. C. Look, B. Claflin, Y. I. Alivov, and S. J. Park, *Phys. Stat. Sol. (a) – Appl. Res.* **2004**, 201, 2203.

15 D. C. Look, *Semicond. Sci. Technol.* **2005**, 20, S55.

16 D. C. Look, R. L. Jones, J. R. Sizelove, N. Y. Garces, N. C. Giles, and L. E. Halliburton, *Phys. Stat. Sol. (a) – Appl. Res.* **2003**, 195, 171.

17 Y. R. Ryu, T. S. Lee, and H. W. White, *Appl. Phys. Lett.* **2003**, 83, 87.

18 A. Tsukazaki, A. Ohtomo, T. Onuma, M. Ohtani, T. Makino, M. Sumiya, K. Ohtani, S. F. Chichibu, S. Fuke, Y. Segawa, H. Ohno, H. Koinuma, and M. Kawasaki, *Nature Mater.* **2005**, 4, 42.

19 Y. R. Ryu, T. S. Lee, J. A. Lubguban, H. W. White, B. J. Kim, Y. S. Park, and C. J. Youn, *Appl. Phys. Lett.* **2006**, 88, Art. No. 241108.

20 H. S. Yang, S. Y. Han, Y. W. Heo, K. H. Baik, D. P. Norton, S. J. Pearton, F. Ren, A. Osinsky, J. W. Dong, B. Hertog, A. M. Dabiran, P. P. Chow, L. Chernyak, T. Steiner, C. J. Kao, and G. C. Chi, *Jpn. J. Appl. Phys. Part 1* **2005**, 44, 7296.

21 J. W. Dong, A. Osinsky, B. Hertog, A. M. Dabiran, P. P. Chow, Y. W. Heo, D. P. Norton, and S. J. Pearton, *J. Electron. Mater.* **2005**, 34, 416.

22 A. Osinsky, J. W. Dong, M. Z. Kauser, B. Hertog, A. M. Dabiran, P. P. Chow, S. J. Pearton, O. Lopatiuk, and L. Chernyak, *Appl. Phys. Lett.* **2004**, 85, 4272.

23 P. Gopal and N. A. Spaldin, *J. Electron. Mater.* **2006**, 35, 538.

24 G. Coli and K. K. Bajaj, *Appl. Phys. Lett.* **2001**, 78, 2861.

25 P. K. Tien and R. Ulrich, *J. Opt. Soc. Am.* **1970**, 60, 1325.

26 M. Born and E. Wolf, *Principles of Optics*, Pergamon, Oxford, 7th Ed. 1999.

27 G. E. Jellison and L. A. Boatner, *Phys. Rev. B.* **1998**, 58, 3586.

28 A. B. Djurisic, Y. Chan, and E. H. Li, *Appl. Phys. A – Mater. Sci. Process.* **2003**, 76, 37.

29 R. Schmidt-Grund, A. Carstens, B. Rheinlander, D. Spemann, H. Hochmut, G. Zimmermann, M. Lorenz, M. Grundmann, C. M. Herzinger, and M. Schubert, *J. Appl. Phys.* **2006**, 99, Art. No. 12371.

30 C. W. Teng, J. F. Muth, U. Ozgur, M. J. Bergmann, H. O. Everitt, A. K. Sharma, C. Jin, and J. Narayan, *Appl. Phys. Lett.* **2000**, 76, 979.

31 R. J. Elliott, *Phys. Rev.* **1957**, 108, 1384.

32 J. F. Muth, R. M. Kolbas, A. K. Sharma, S. Oktyabrsky, and J. Narayan, *J. Appl. Phys.* **1999**, 85, 7884.

33 W. Franz, *Z. Naturforsch. a* **1958**, 13, 484.

34 L. V. Keldysh, *Sov. Phys. JETP* **1958**, 7, 788.

35 R. J. Elliott, *Phys. Rev.* **1957**, 108, 1384.

36 J. D. Dow and D. Redfield, *Phys. Rev. B.* **1970**, 1, 3358.

37 J. D. Dow and D. Redfield, *Phys. Rev. B.* **1972**, 5, 594.

38 M. Wraback, H. Shen, S. Liang, C. R. Goria, and Y. Lu, *Appl. Phys. Lett.* **1999**, 74, 507.

39 A. E. Oberhofer, J. F. Muth, M. A. L. Johnson, Z. Y. Chen, E. F. Fleet, and G. D. Cooper, *Appl. Phys. Lett.* **2003**, 83, 2748.

40 A. Ohtomo, M. Kawasaki, T. Koida, K. Masubuchi, H. Koinuma, Y. Sakurai, Y. Yoshida, T. Yasuda, and Y. Segawa, *Appl. Phys. Lett.* **1998**, 72, 2466.

41 A. K. Sharma, J. Narayan, J. F. Muth, C. W. Teng, C. Jin, A. Kvit, R. M. Kolbas, and O. W. Holland, *Appl. Phys. Lett.* **1999**, 75, 3327.

42 Y. R. Ryu, T. S. Lee, J. A. Lubguban, A. B. Corman, H. W. White, J. H. Leem, M. S. Han, Y. S. Park, C. J. Youn, and W. J. Kim, *Appl. Phys. Lett.* **2006**, 88, Art. No. 052103.

43 K. Sakurai, T. Kubo, D. Kajita, T. Tanabe, H. Takasu, and S. Fujita, *Jpn. J. Appl. Phys. Part 2* **2000**, 39, L1146.

44 T. Gruber, C. Kirchner, R. Kling, F. Reuss, and A. Waag, *Appl. Phys. Lett.* **2004**, 84, 5359.

45 W. J. Kim, J. H. Leem, M. S. Han, Y. R. Ryu, and T. S. Lee, *J. Appl. Phys.* **2006**, 99, Art. No. 096104.

46 M. E. Kolanz, *Appl. Occupational Environm. Hygiene,* **2001**, 16, 559.

47 K. Sakurai, T. Takagi, T. Kubo, D. Kajita, T. Tanabe, H. Takasu, and S. Fujita, *J. Crystal Growth* **2002**, 237, 514.

48 T. Makino, Y. Segawa, M. Kawasaki, A. Ohtomo, R. Shiroki, K. Tamura, T. Yasuda, and H. Koinuma, *Appl. Phys. Lett.* **2001**, 78, 1237.

49 J. Ishihara, A. Nakamura, S. Shigemori, T. Aoki, and J. Temmyo, *Appl. Phys. Lett.* **2006**, 89, Art. No. 091914.

50 A. V. Osinsky, J. W. Dong, J. Q. Xie, B. Hertog, A. M. Dabiran, P. P. Chow, S. J. Pearton, D. P. Norton, D. C. Look, W. Schoenfeld, O. Lopatiuk, L. Chernyak, M. Cheung, A. N. Cartwright, and M. Gerhold, *2005 Mater. Res. Soc. Fall Meeting, Proceedings* **2005**, Paper EE9/FF18.

51 "Progress made in ZnO emitters", *Laser Focus World* **2006**, January.

Index

Wide Bandgap Light Emitting Materials and Devices. Edited by G. F. Neumark, I. L. Kuskovsky, and H. Jiang
Copyright © 2007 WILEY-VCH Verlag GmbH & Co. KGaA, Weinheim
ISBN: 978-3-527-40331-8

Related Titles

Piprek, J. (ed.)

Nitride Semiconductor Devices: Principles and Simulation

2007. Hardcover
ISBN 978-3-527-40667-8

Deveaud, B. (ed.)

The Physics of Semiconductor Microcavities

From Fundamentals to Nanoscale Devices

2007. Hardcover
ISBN 978-3-527-40561-9

Lanzani, G. (ed.)

Photophysics of Molecular Materials

From Single Molecules to Single Crystals

2006. Hardcover
ISBN 978-3-527-40456-8

Rincon-Mora, G. A.

Voltage References

From Diodes to Precision High-Order Bandgap Circuits

2002. Softcover
ISBN 978-0-471-14336-9

Yeh, P., Gu, C.

Optics of Liquid Crystal Displays

1999. Hardcover
ISBN 978-0-471-18201-6

Fukuda, M.

Optical Semiconductor Devices

1999. Hardcover
ISBN 978-0-471-14959-0